Durability of Reinforced Concrete Structures

Durability of Reinforced Concrete Structures

Paul Chess and Warren Green

CRC Press
Taylor & Francis Group
Boca Raton London New York

CRC Press is an imprint of the
Taylor & Francis Group, an **informa** business

CRC Press
Taylor & Francis Group
6000 Broken Sound Parkway NW, Suite 300
Boca Raton, FL 33487-2742

First issued in paperback 2021

ISBN-13: 978-0-367-27838-0 (hbk)
ISBN-13: 978-1-03-217608-6 (pbk)
DOI: 10.1201/9780429298189

Library of Congress Cataloging-in-Publication Data

Names: Chess, Paul, author. | Green, Warren (Corrosion engineer), author.
Title: Durability of reinforced concrete structures / by Paul Chess and Warren Green.
Description: First edition. | Boca Raton, FL : CRC Press/Taylor & Francis Group, [2020] | Includes bibliographical references and index. |
Summary: "Reinforced concrete structures corrode as they age, with significant financial implications, but it is not immediately clear why some are more durable than others. This book looks at the mechanisms for corrosion and how corrosion engineering can be used for these problems to be minimized in future projects. Several different examples of reinforced concrete structures with corrosion problems are described and the various life enhancement solutions considered and applied are discussed. The book includes a chapter on the effectiveness of corrosion monitoring techniques and questions why the reality is at odds with current theory and standards. Specialist contractors, consultants and owners of corrosion damaged structures will find this an extremely useful resource. It will also be a valuable reference for students at postgraduate level"-- Provided by publisher.
Identifiers: LCCN 2019043459 (print) | LCCN 2019043460 (ebook) | ISBN 9780367278380 (hardback ; acid-free paper) | ISBN 9780429298189 (ebook)
Subjects: LCSH: Reinforced concrete construction. | High strength concrete. | Concrete--Corrosion.
Classification: LCC TA440 .C524 2020 (print) | LCC TA440 (ebook) | DDC 624.1/8341--dc23
LC record available at https://lccn.loc.gov/2019043459
LC ebook record available at https://lccn.loc.gov/2019043460

Visit the Taylor & Francis Web site at
http://www.taylorandfrancis.com

and the CRC Press Web site at
http://www.crcpress.com

Contents

Authors

Paul Chess is currently the Managing Director of Corrosion Remediation Limited. He was formerly the Managing Director of the largest specialist manufacturer of products for cathodic protection of concrete in the world.

Warren Green is a Director and a Principal Corrosion Engineer of Vinsi Partners, based in Sydney, Australia. He is also an Adjunct Associate Professor in the Institute for Frontier Materials at Deakin University, Victoria, Australia, and a Visiting Adjunct Associate Professor within the Curtin Corrosion Centre of the Western Australia School of Mines: Minerals, Energy and Chemical Engineering (MECE) at Curtin University, Perth, Australia.

Authors

The Problem

1

1.1 REINFORCED CONCRETE

Imagine the world without steel-reinforced concrete structures and buildings. Imagine our primitive existence without the wonder and delight of steel-reinforced concrete.

Concrete itself is not a modern material. It has been in use for thousands of years. Calcined impure gypsum was used as a cementing material by the ancient Egyptians (Neville, 2011). Since Roman times lime-based hydraulic cements (those set by chemical reaction with water) have been in use and such ancient mortars still survive today, including examples in the Pantheon in Rome and aqueducts in France. The word concrete comes from the Latin *concretus*, which means 'mixed together' or compounded (Macdonald, 2003).

The Romans ground together lime and volcanic ash or finely ground burnt clay tiles. The active silica and alumina in the ash or tiles combined with the lime to produce what became known as 'pozzolanic cement' from the name of the village of Pozzuoli, near Vesuvius, where the volcanic ash was first found (Neville, 2011). The name 'pozzolanic cement' is used to this day to describe cements obtained simply by the grinding of natural materials at normal temperature (Neville, 2011).

The Middle Ages brought a decline in the use of hydraulic mortars until the 18th century when John Smeaton in 1756 used a mortar produced from pozzolana mixed with a limestone containing a considerable proportion of clayey material to rebuild Eddystone Lighthouse, off the Cornish coast of England (Neville, 2011).

Lime concrete was the precursor to modern concrete and was made with natural hydraulic cements (Macdonald, 2003). The first modern concrete was made with Portland cement as patented by Joseph Aspdin in 1824, a Leeds bricklayer, stonemason and builder (Neville, 2011). The name 'Portland cement' is because of its resemblance to stone quarried near Portland in Dorset, England.

The name 'Portland cement' has remained throughout the world to this day to describe a cement obtained by mixing together calcareous and argillaceous,

TABLE 1.1 Properties of Steel and Concrete

PROPERTY	CONCRETE	STEEL
Strength in tension	Poor	Good
Strength in compression	Good	Buckling can occur
Strength in shear	Fair	Good
Durability	Good	Corrodes if unprotected
Fire resistance	Good	Poor, low strength at high temperatures

or other silica-, alumina- and iron oxide-bearing materials in appropriate proportions, burning them at clinkering temperature (around 1,500°C), cooling, before grinding with about 5% gypsum (calcium sulphate).

Concrete then basically consists of mineral aggregate held together by a cement paste, so if we consider a normal mix it will consist of cement, sand (fine aggregate), coarse aggregate and water (and often other admixtures) which are mixed together to form eventually a hard, strong material.

The principles of reinforcing concrete to provide both tensile and compressive strength were understood in ancient times. However, it was not until the 19th century that a number of European and North American inventors developed and patented reinforcing methods. This resulted in the widespread introduction of a fully fledged concrete industry (Macdonald, 2003).

Concrete is strong in compression but weak in tension and shear and so much of the concrete is reinforced, usually with steel. The steel reinforcement can take the form of conventional carbon steel (black steel), prestressing steel, post-tensioned steel and steel fibres, and its widespread utility is primarily due to the fact it combines the best features of concrete and steel. The properties of these two materials are compared in Table 1.1.

Since its inception in the mid-19th century, steel-reinforced concrete has become the most widely used construction material in the world (and the second most-used material by mankind after water). Reinforced concrete is a wonderful composite material. It combines the best features of concrete and steel. Concrete gives it strength in compression while the steel gives it strength in tension and shear.

1.2 STEEL REINFORCEMENT

Rebar (short for reinforcing bar), collectively known as reinforcing steel, reinforcement steel or steel reinforcement, is a steel bar or mesh of steel wires.

The most common type of rebar is carbon steel.

There is also other steel reinforcement including: galvanised steel reinforcement; stainless steel reinforcement; and metallic-clad steel reinforcement for use in reinforced concrete construction.

Reinforced concrete construction using other alternate reinforcement such as bronze, fibre-reinforced polymer (FRP) reinforcement or basalt, is also available in limited quantities.

1.3 PRESTRESSED AND POST-TENSIONED CONCRETE

Prestressed concrete is composed of high-strength concrete and high-strength steel. The concrete is 'prestressed' by being placed under compression by the tensioning of high-strength 'steel tendons' located within or adjacent to the concrete volume.

Prestressed concrete is used in a wide range of buildings and civil structures where its improved performance can allow longer spans, reduced structural thicknesses, and material savings compared to conventionally reinforced concrete. Applications can include high-rise buildings, residential buildings, foundation systems, bridge and dam structures, silos and tanks, industrial pavements and nuclear containment structures.

Prestressed concrete members can be divided into two basic types: pre-tensioned and post-tensioned.

Pre-tensioned concrete is where the high-strength steel tendons are tensioned *prior* to the surrounding concrete being cast. The concrete bonds to the tendons as it cures, following which the end-anchoring of the tendons is released, and the tendon tension forces are transferred to the concrete as compression by static friction.

Post-tensioned concrete is where the high-strength steel tendons are tensioned *after* the surrounding concrete has been cast. The tendons are not placed in direct contact with the concrete, but are encapsulated within a protective duct typically constructed from plastic or galvanised steel materials. There are then two main types of tendon encapsulation systems: those where the tendons are subsequently bonded to the surrounding concrete by internal grouting of the duct after stressing (*bonded* post-tensioning); and those where the tendons are permanently de-bonded from the surrounding concrete, usually by means of a greased sheath over the tendon strands (*unbonded* post-tensioning).

1.4 CONCERNS ABOUT DURABILITY OF REINFORCED CONCRETE

The environments that we place our reinforced concrete structures in mean that various deterioration processes can affect their in-service durability, leading to loss of functionality, unplanned maintenance/remediation/replacement, and in the worst cases, loss of structural integrity and resultant safety risks.

Deterioration of concrete can be separated into two broad category types: (i) degradation of concrete, and (ii) corrosion of steel reinforcement. Figure 1.1 summarises the causes of (i) and (ii) which can include one or more or mechanical, physical, structural, chemical, biological and reinforcement corrosion mechanisms.

Amongst these, the most common cause of deterioration of reinforced concrete structures is corrosion of conventional carbon steel (black steel), prestressing steel and post-tensioned steel reinforcement.

1.5 CORROSION OF STEEL REINFORCEMENT IN CONCRETE

1.5.1 Causes

The passivity (thin protective oxide film) provided to steel reinforcement by the alkaline environment of concrete may be lost or locally compromised if

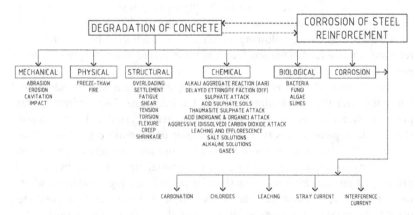

FIGURE 1.1 Summary of concrete deterioration types. (Adapted from Bertolini et al., 2004.)

the pH of the concrete pore solution falls because of carbonation or if aggressive ions such as chlorides penetrate in sufficient concentrations to the steel reinforcement surface.

Carbonation of concrete occurs as a result of atmospheric CO_2 gas (and atmospheric SO_x and NO_x gases) neutralising the concrete pore water (lowering its pH to 9) and thereby destroying (dissolving) the passive film.

Chloride ions penetrate concrete principally by concentration-driven diffusion but also with the movement of moisture. When the chloride ions reach the reinforcement surface in sufficient concentrations, they 'punch holes' in the passive film leading to localised (pitting) corrosion of the steel.

Leaching of $Ca(OH)_2$ (and NaOH and KOH) from concrete also lowers pH to cause corrosion of steel reinforcement.

Stray electrical currents, most commonly from electrified traction systems, can also break down the passive film and cause corrosion of steel-reinforced and prestressed concrete elements. This is particularly a problem with DC systems.

Interference currents, typically associated with cathodic protection systems for large steel objects such as marine jetties, can also break down the passive film and cause corrosion of steel reinforced and prestressed concrete elements.

1.5.2 Uniform (Microcell) Corrosion and Pitting (Macrocell) Corrosion

According to the different spatial location of anodes and cathodes, corrosion of steel in concrete can occur in different forms (Elsener, 2002):

- As *microcells*, where anodic and cathodic reactions are immediately adjacent, leading to uniform steel (iron) dissolution over the whole surface. This uniform (or general) corrosion is typically caused by carbonation of the concrete or by very high chloride content at the steel reinforcement.
- As *macrocells*, where a net distinction between corroding areas of the steel reinforcement (anodes) and non-corroding passive surfaces (cathodes) is found. Macrocells occur mainly in the case of chloride-induced corrosion (pitting) where the anodes are small with respect to the total (passive) steel reinforcement surface.

It is suggested (Cherry and Green, 2017) that uniform or general corrosion of steel reinforcement in concrete where the cathodic and anodic sites may be separated by millimetres (rather than microns), can be termed *minicell* corrosion.

1.5.3 Corrosion Products Development – Cracking, Delamination and Spalling

The process of reinforcement corrosion will lead to corrosion products which, generally speaking, will occupy a greater volume than the iron dissolved in its production, see Figure 1.2 (Jaffer and Hanson, 2009). Furthermore, when the corrosion products become hydrated the volume increase is even greater (Broomfield, 2007): see hydrated haematite (α-$Fe_2O_3.3H_2O$) (red rust) in Figure 1.2.

The consequence of higher volume corrosion products is then to develop tensile stresses in the concrete covering the reinforcement. Concrete, being weak in tension, will crack as a consequence of the corrosion. Continuing formation of corrosion product(s) will enhance the expansion which will ultimately lead to cracked pieces of concrete cover detaching, leading to delamination and then spalling. Rust staining of the concrete may or may not occur together with the cracking or as a prelude to delamination and spalling. Section loss, bond loss and anchorage loss of the reinforcement also occurs as a result of the corrosion.

Hydrated haematite (α-$Fe_2O_3.3H_2O$) (red rust)-induced cracking, delamination and spalling of cover concrete can be frequently seen in deteriorated

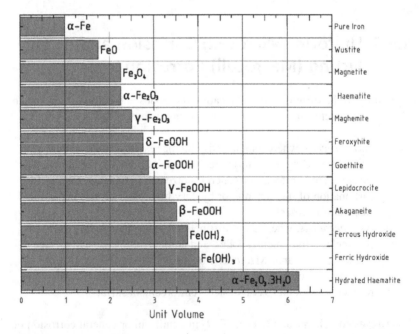

FIGURE 1.2 Relative volume of iron and some of its oxides. (From Jaffer and Hansson, 2009.)

reinforced concrete structures, an example of which is provided at Figure 1.3 for a marine structure.

It is most important to note, however, that not all corrosion of reinforcement leads to rust staining, cracking, delamination or spalling of cover concrete. Pitting, localised corrosion at cracks and localised corrosion at concrete defects, can result in marked section loss (loss of bond, loss of anchorage) and ultimately structural failure without the visible consequences of corrosion on the concrete surface, i.e. no rust staining, cracking, delamination or spalling of cover concrete (Green et al., 2013). An example of marked localised section loss of reinforcement due to chloride-induced pitting (macrocell corrosion) is provided at Figure 1.4. The definition of macrocell corrosion is given in Section 1.5.2 where a net distinction between corroding areas of the steel reinforcement (anodes) and non-corroding passive surfaces (cathodes) is found.

Elsener (2002) pointed out that macrocell corrosion (pitting) is of great concern because the local dissolution rate (reduction in cross-section of the conventional/prestressing/post-tensioned steel reinforcement, loss of bond, loss of anchorage) may greatly be accelerated due to the large cathode/small anode area ratio. Indeed, values of local corrosion (penetration) rates of up to 1 mm/year have been reported for bridge decks and other chloride-contaminated steel-reinforced concrete structures, according to Elsener (2002).

Broomfield (2007) also notes that 'black rust' or 'green rust' (due to the colour of the liquid when first exposed to air after breakout) corrosion can also be found under damaged waterproof membranes and in some underwater or water-saturated structures. He states that it is potentially dangerous as there is no indication of corrosion by cracking and spalling of the concrete and

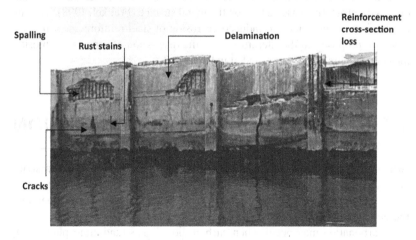

FIGURE 1.3   Reinforcement corrosion-induced deterioration to a marine structure.

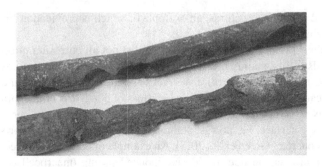

FIGURE 1.4 Example of marked pitting (macrocell corrosion) and section loss (corrosion penetration) of steel reinforcement. Here pits would have started out narrow but with time have coalesced to result in section loss over a greater (anodic) area.

the reinforcing steel may be severely weakened before corrosion is detected. Reinforcement bars may be 'hollowed out' in such deoxygenated conditions.

1.6 CHLORIDE-INDUCED CORROSION

It is known that chloride ions in sufficient concentration can destroy the passivity of steel reinforcement (conventional, prestressed and post-tensioned) in concrete, leading to pitting (macrocell corrosion). Pitting is localised accelerated dissolution of a metal that occurs as a result of the breakdown of the otherwise protective passive film on the metal surface (Frankel, 1998). Various mechanisms of chloride ion-induced corrosion of steel reinforcement in concrete are proposed in the literature. Specific discussion of aspects of chloride ion-induced corrosion are provided in Chapter 2.

1.7 CARBONATION-INDUCED CORROSION

Attack of concrete by carbon dioxide is termed carbonation. The passivity provided to steel reinforcement by the alkaline environment of concrete can be lost due to a fall in the pH of the concrete pore solution because of carbonation, thereby destroying the passive film.

Carbonation may occur when carbon dioxide gas (and atmospheric SO_x and NO_x gases) from the atmosphere dissolves in concrete pore water and penetrates inwards or when the concrete surface is exposed to water or soil containing dissolved carbon dioxide.

Carbon dioxide dissolves in the pore water to form carbonic acid by the reaction:

$$CO_2(g) + H_2O(l) \rightarrow H_2CO_3(aq) \tag{1.1}$$

Carbonic acid can dissociate into hydrogen and bicarbonate ions. The carbonic acid reacts with $Ca(OH)_2$ (portlandite) in the solution contained within the pores of the hardened cement paste to form neutral insoluble $CaCO_3$. The general reaction is as follows:

$$Ca(OH)_2(aq) + H_2CO_3(aq) \rightarrow CaCO_3(s) + 2H_2O(l) \tag{1.2}$$

and the net effect is to reduce the alkalinity of the pore water which is essential to the maintenance of a passive film on any reinforcing steel that may be present. Whilst there is a pH buffer between CaO and $Ca(OH)_2$ that keeps the pH at approximately 12.5, the CO_2 keeps reacting until the buffer is consumed and then the pH will drop to levels no longer protective of steel.

The attack of buried concrete by carbon dioxide dissolved in the groundwater is a two-stage process. The calcium hydroxide solution that fills the pores of the concrete first reacts with dissolved carbon dioxide to form insoluble calcium carbonate. However, it subsequently reacts with further carbon dioxide to form soluble calcium bicarbonate which is leached from the concrete. The extent to which each process takes place is a function of the calcium carbonate/calcium bicarbonate concentration of the groundwater (which in turn is a function of the pH and the calcium content) and the amount of dissolved carbon dioxide.

Carbonation of concrete causes general corrosion where anodic and cathodic reactions are immediately adjacent (microcells or minicells), leading to uniform steel (iron) dissolution over the whole surface.

Localised general (uniform) carbonation-induced corrosion of steel reinforcement can occur at cracks, concrete defects, low cover areas, etc. of reinforced concrete elements.

1.8 LEACHING-INDUCED CORROSION

The passivity provided to steel reinforcement by the alkaline environment of concrete may also be lost if the pH of the concrete pore solution falls because of leaching of $Ca(OH)_2$ (and NaOH and KOH). Leaching from concrete can lead to a lowering of the pH below 10 to cause corrosion of steel reinforcement.

Natural waters may be classified as 'hard' or 'soft' usually dependent upon the concentration of calcium bicarbonate that they contain. Hard waters may contain a calcium ion content in excess of 10 mg/l, whereas a soft water may contain less than 1 mg/l calcium. The capillary pores in a hardened cement paste contain a saturated solution of calcium hydroxide which is in equilibrium with the calcium silicate and aluminate hydrates that form the cement gel. If soft water can penetrate through the concrete (e.g. joint, crack, defect, etc.) then it can leach free calcium hydroxide out of the hardened cement gel so that the pore water is diluted and the pH falls. Since the stability of the calcium silicates, aluminates and ferrites that constitute the hardened cement gel requires a·certain concentration of calcium hydroxide in the pore water, leaching by soft water can result in decomposition of these hydration products. The removal of the free lime in the capillary solution leads to dissolution of the calcium silicates, aluminates and ferrites, and this hydrolytic action can continue until a large proportion of the calcium hydroxide is leached out, leaving the concrete with negligible strength. Prior to this stage, loss of alkalinity due to the leaching process will result in reduced corrosion protection to reinforcement.

Calcium hydroxide that is leached to the concrete surface reacts with atmospheric carbon dioxide to form deposits of white calcium carbonate on the surface of the concrete. These deposits of calcium carbonate may take the form of stalactites or severe efflorescence.

Leaching-induced corrosion of steel reinforcement can be localised and consequently more serious where it occurs at concrete joints, cracks, defects, etc. An example of localised corrosion of steel reinforcement in a potable water containing reinforced concrete tank is shown at Figure 1.5.

1.9 STRAY AND INTERFERENCE CURRENT-INDUCED CORROSION

1.9.1 General

Stray current causing increased rates of corrosion is a significant problem for steel structures in the ground or water. This current can come from many sources such as train traction systems, leakage from high voltage transmission lines, cathodic protection installations, to name but a few and can be of the

FIGURE 1.5 'Black rust' (magnetite) localised corrosion of steel reinforcement due to leaching at construction joints in the wall of a reinforced concrete tank containing potable water. (Courtesy of Greg Moore.)

AC or DC variety. The assumption has been (ISO/DIS 21857, 2019) that that the same damage mechanism will occur as steel in soil or water but is this an incorrect assumption? This is because firstly, the concrete coating (cover) over the steel will dramatically increase the resistance of the likely corrosion circuit and secondly, the chemical reaction of the concrete at the anodic corrosion site will prevent the dissolution of the steel.

When testing of reinforced concrete specimens has actually been conducted it has been found that if the concrete does not include any chloride (apart from permissible background levels), then anodic current densities of $1A/m^2$ steel reinforcement area could not initiate corrosion even after being applied for 14 months. This is in contrast to active corrosion at $10mA/m^2$ and higher current densities for steel not protected by the concrete in the same trial. It was found possible to initiate corrosion at $10A/m^2$ as the acidification caused by this huge current overwhelmed the concrete's alkalinity and counter-diffusion process. These results in practice mean that reinforced concrete is immune to DC-induced corrosion in non-chloride-containing groundwaters and soils unless it is directly beside a very large leakage current such as a traction system.

The AC corrosion risk is substantially lower than the DC risk, so again there is no concern practically with this form of degradation.

Where things are different is when there is chloride in the concrete or in the water or soil outside the concrete. The presence of chloride at the steel reinforcement surface massively reduces the amount of stray corrosion current required to initiate corrosion. Research (Bertolini et al., 2013) has shown that the presence of 0.2% (by weight of chloride) reduced by a factor of 100 the stray current density required to initiate corrosion. The presence of the stray current circuit will also move chlorides from the outside of the concrete to the steel reinforcement by electro-migration. This mechanism will rapidly increase the concentration of the chloride at the steel reinforcement in a few weeks at a significant current density.

A reinforced concrete structure which has suffered high levels of degradation is prestressed concrete cylinder pipe (PCCP). This has an American origin and is widely used throughout the world for carrying fresh or waste water. The problem occurs when laying these pipelines in salt contaminated soil. Because of the difference in grain size between the soil and the concrete matrix there will be an inward flow of ions with the chloride advected into the concrete (see Chapter 2). Combine this with telluric effects as the pipelines are long and severe damage can, and does, occur prematurely in isolated sections where the post-tensioned wire is attacked. This commonly leads to pipe failure which is very expensive to repair.

In summary the ground environment is critical in judging the susceptibility of reinforced concrete to stray current corrosion.

1.9.2 Ground Currents

Ground currents can cause significant corrosion in below grade reinforced concrete structures such as concrete pipelines, tunnels and retaining walls which are in a saline environment because there can be large and sustained current flows

The commonest cause of stray current corrosion is electrified traction systems which after a few years' operation have a tendency to spew out current over any metallic item in the close to middle environment. The DC power for an electric train or tram is supplied by a sub-station and drawn from a positive overhead cable or one of the rails. The negative return to the sub-station is either through another rail on which the train is running or through a separate conductor rail. Although most of the current driving the trains/trams returns to the substation from which it is drawn via the conductor rail of the traction system, it is inevitable that some of the current in the conductor rail will leak from that rail and go back to the sub-station through the earth. This sets up potential gradients in the earth. If a low resistance path (such as metallic reinforcement in a concrete structure) is present then the current may pass through the reinforcement. It can be seen from Figure 1.6 that where the current enters the structure, the reinforcement at this location is cathodically protected. Where the current leaves the structure near the sub-station, corrosion may occur in certain circumstances.

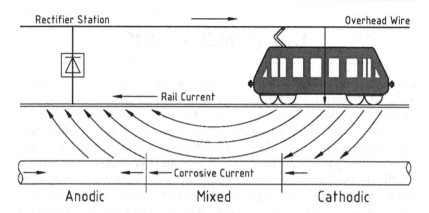

FIGURE 1.6 Ground currents set up by electric traction system. (From Cherry and Green, 2017.)

1.9.3 Interference Currents

Interference corrosion currents are typically associated with cathodic protection schemes. The application of remote anode impressed current cathodic protection to (say) a buried pipeline involves the establishment of anodes (anode groundbed) at some distance from the pipeline with the aim of passing current from the anodes (groundbed) to the pipeline. If there is a reinforced concrete structure within this potential field then potential gradients associated with the passage of current through the soil could possibly cause interference corrosion of steel reinforcement (though this is unlikely considering the current output of a coated pipeline CP system). Normally it is only CP systems on steel jetties which are big enough to cause potential problems with the concrete in seawater

1.9.4 Local Corrosion Due to Stray or Interference Currents

If the reinforced concrete structure is dumping current at a particular location the corrosion of steel reinforcement will be localised. For most reinforced concrete structures this will not cause a structural problem even in a saline environment, but when the steel is post-tensioned (here the tendon has to be in electrical connection through the duct which is rare) or prestressed this may be critical.

1.10 CONCLUSIONS

Concrete is a most wonderful material of construction. When suitably designed, constructed and maintained, reinforced concrete provides service lives of numerous decades to structures and buildings. Concrete provides reinforcing steel with excellent corrosion protection. The highly alkaline environment in concrete results in the spontaneous formation of a stable, tightly adherent, thin protective oxide passive film on the steel reinforcement surface, which protects it from corrosion. In addition, well-proportioned, compacted and cured concrete has a low penetrability, thereby minimising the ingress of corrosion-inducing species via the aqueous phase, and a relatively high electrical resistivity, which reduces the corrosion current and hence the rate of corrosion if corrosion is initiated.

There are however, several degradative processes which affect some reinforced concrete structures leading to loss of functionality, unplanned maintenance/remediation/replacement, and in the worst cases, loss of structural integrity and resultant safety risks. Amongst these, the most common cause of deterioration is corrosion of conventional carbon steel (black steel), prestressing steel and post-tensioned steel reinforcement.

REFERENCES

Bertolini L, Elsener B, Pedeferri P, Redaelli E and Polder R B (2013), *Corrosion of Steel in Concrete: Prevention, Diagnosis, Repair*, Second Edition, Wiley-VCH Verlag GMbH & Co KGaA, Weinheim.

Broomfield J P (2007), *Corrosion of Steel in Concrete*, Second Edition, Taylor and Francis, Oxon.

Cherry B and Green W (2017), The Corrosion and Protection of Concrete Structures and Buildings, Australasian Corrosion Association and Australasian Concrete Repair Association, Training Course Notes, Version 1.0, Melbourne, Australia.

Elsener B (2002), Macrocell corrosion of steel in concrete – implications for corrosion monitoring, *Cement & Concrete Composites*, 24, 65–72.

Frankel G S (1998), Pitting corrosion of metals a review of the critical factors, *Journal of the Electrochemical Society*, 145, 6, 2186–2198.

Green W, Dockrill B and Eliasson B (2013), Concrete repair and protection – overlooked issues, Proceedings of Corrosion and Prevention 2013 Conference, Australasian Corrosion Association Inc., Brisbane, November, Paper 020.

International Standards Association (2019), ISO/DIS 21857 Petroleum, petrochemical and natural gas industries - Prevention of corrosion on pipeline systems influenced by stray currents.

Jaffer S J and Hansson C M (2009), Chloride-induced corrosion products of steel in cracked-concrete subjected to different loading conditions, *Cement and Concrete Research*, 29, 116–125.

Macdonald S (2003), *The Investigation and Repair of Historic Concrete*, NSW Heritage Office, Parramatta, New South Wales, Australia.

Neville A M (2011), *Properties of Concrete*, Fifth Edition, Pearson Education Limited, Essex, England.

The Corrosion Process in Reinforced Concrete: The State of the Art

2

2.1 INTRODUCTION

The biggest durability problem with structures comprised of reinforced concrete is corrosion of the steel reinforcement. The most likely cause of this corrosion, which is also the most difficult to remedy, is the presence of chloride in the concrete. This chloride effectively catalyses the steel corrosion process. Because of the huge economic significance of this problem over the last 50 years there has been almost continuous research in many countries to define and categorise the problem. A lot of the research has been replicated in different countries and by particular research bodies. One of the biggest themes of this research has been to provide a chloride concentration in steel at which a significant level of corrosion occurs. This defined level is a cornerstone of most civil engineering codes and this chapter looks at why this is a gross oversimplification of the actual situation. Most literature considers the movement of ions through concrete to occur in the same manner as in an aqueous solution by concentration diffusion. In this chapter the evidence that electromigration is an important mover of certain cation and anion species is developed and also the evidence that advection can be an extremely significant transport mechanism.

2.2 THE CORROSION PROCESS

How the steel corrodes in concrete is of significant importance in trying to understand the likely mechanisms and possible rates. If a steel specimen is grit-blasted to a bright metal finish and then put in a saline, pH neutral, aqueous solution then there will be a rapid change over the whole surface area. A bright yellowy brown oxide will form within minutes and gradually become darker as the exposure time increases. This is commonly referred to as microcell corrosion as it appears to the naked eye to be evenly distributed over the whole surface of the rebar. As the exposure time increases the rate of corrosion reduces as the oxide layer provides a barrier to ionic, atomic and gaseous transport.

In reinforced concrete structures examined when the contaminated concrete cover is removed after many years' exposure, the underlying rebar often shows differing corrosion products depending on the rebar's spatial orientation. There can be areas where the corrosion appears to be general with uniform section loss (Figure 2.1). This tends to be in areas with lower cover depth, dry and carbonated concrete but this is not universally true.

There can also be corrosion pitting where there is a significant loss of section at defined locations. This is most commonly found at cracks, concrete defects and in wet and high cover parts of a structure. Various forms of pitting have been observed, as shown in Figure 2.2. The most commonly observed of

FIGURE 2.1 Example of significant general (uniform) corrosion.

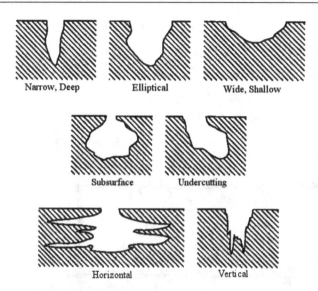

FIGURE 2.2   Examples of pitting corrosion forms.

these forms is the elliptical shape in conventionally reinforced and prestressed concrete.

When exposed the material found in these pits often has some black, green or yellow colourations and can smell of hypochlorous acid, recognisable from swimming baths. Sometimes, but rarely, the corrosion pits can be bare and this is where there appears to be flushing of the concrete, such as by tidal action. In both of these cases the oxide films (corrosion products) do not provide a significant protective barrier to slow the rate of corrosion. In fact, the corrosion products may be providing a poultice where a low pH and aggressive ions can remain in contact with the metal, with the result that the corrosion rate is actually increased. The reason for this difference in behaviour between general corrosion and pitting corrosion is complex with both the morphology of the corrosion product and the composition of the immediate concrete interface responsible.

There are at least two distinct corrosion types, with many forms of corrosion product. This makes categorising the corrosion reaction as a simple one-stage process untenable. Another aspect of the corrosion cell which can also be distinctly different is that you can have the complete corrosion cell in under a millimeter or it can be several centimeters or more, as shown schematically in Figure 2.3. The size of this complete cell has ramifications for the ionic corrosion mechanism and ionic transport. Several researchers (Green et al., 2017) have looked at different hypotheses for pit formation and presently the most favoured is the so-called 'transitory complex model' where the chloride ions

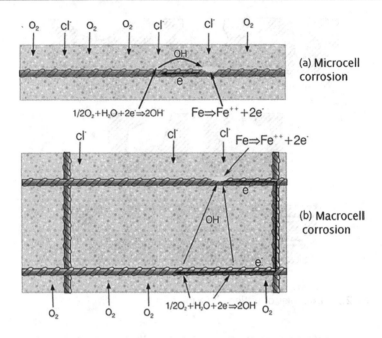

FIGURE 2.3   Schematic of micro- and macrocell corrosion mechanisms.

form a soluble compound with iron ions which then moves away from the anodic site at the base of the pit. At a position remote from the pit, which is the initial corrosion site, a compound identified as iron hydroxide precipitates releasing the chloride ions. These chloride ions then return to the pit. This mechanism has been observed (Angst et al., 2011) on several reinforced concrete specimens with initial pit formation and the initial corrosion product deposited on the rebar several millimeters away from the pit, as is shown schematically in Figure 2.4.

There are several stages to pit formation and it is likely that the actual mechanisms will be different, as there is not a simple direct corrosion process but a multi-stage chain of reactions which will vary depending on many factors, including ionic transport. What the above findings demonstrate is that the corrosion process of steel in reinforced concrete is much more complex than that normally

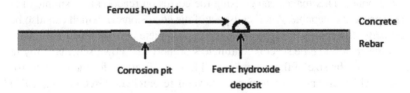

FIGURE 2.4   Schematic of pit transport process.

represented in an idealised solution and thus cannot be adequately modelled using the standard electrolytic formulas such as the Stern-Geary equation. This is an equation that relates quantitatively the slope of a polarisation curve in the vicinity of the natural corrosion potential to the corrosion current density.

One of the many variables in the corrosion of steel in concrete which has only recently started to be explored are the differences in the composition and structure of the reinforcement itself having a significant effect on corrosion initiation and propagation. For example, it has been found that manganese sulphide inclusions can preferentially be corroded in chloride-rich reinforced concrete and this could initiate or fully form pits. Other observations have discerned that the technique used for the manufacture of the rebar also has a significant influence on the form and extent of the corrosion process. Recently the entire steel-concrete interface has been looked at closely and a review published (Geiker et al., 2017).

The initial iron oxidation reaction is an anodic process and involves the loss of electrons. This initial reaction is then followed by several further reactions leading to a more ionically charged atom and further electron production. The salient point here is that there are electrons being produced and that these are travelling through the steel matrix to the cathode on the steel surface, which is a non-corroding area where they combine with oxygen and water in the cathodic reduction reaction. When looking at heavily corroded reinforcement, almost all of it is has section loss, suggesting that this cathode is a significant distance away. Figure 2.5 is an example of corrosion of reinforcing steel causing spalling concrete and replacement of the bridge.

FIGURE 2.5 Corrosion of reinforcing steel causing spalling concrete and replacement of the bridge.

In microcell-type corrosion the anode and cathode can be part of the same grain of steel which is typically from 0.25 mm down to 0.025 mm in size. This is because steel is an alloy with component parts, such as iron carbide and ferrite. These have different chemical compositions and this means they will have different electrical potentials promoting the formation of small anodes and cathodes in the individual grains. In macrocell corrosion the difference in potential of these microcell anodes is outweighed by a larger potential difference making this area all a cathode or anode.

2.3 THE AMOUNT OF CHLORIDE REQUIRED TO INITIATE CORROSION

This has been exhaustively evaluated over many years and a review of the many experimental procedures undertaken was published (Angst et al., 2009). In this review the parameters which affect the onset and amount of corrosion were discussed and found to include 16 variables.

With all these variables it is perhaps not surprising that there would be a very significant range in the critical chloride level at which corrosion is initiated. For laboratory studies these ranged from 0.04–8.4% chloride by weight of cement which is a 21,000% difference (Angst et al., 2009). This huge difference was also found in real structures where the range was from 0.1% to 1.95% chloride by weight of cement which is a 1,950% difference. What this means is that the presence of chloride at any practically measurable level may initiate corrosion but the presence of large amounts of chloride does not necessarily mean that corrosion will occur and it is other factors which determine whether this initiation and propagation phase occur.

2.4 THE EFFECT OF THE STRUCTURE'S SHAPE ON CORROSION

Recently an experiment was undertaken (Angst and Elsener, 2017) where the specimen size was varied in the same experimental procedure. It was found with exactly the same experimental conditions and procedures that reinforced concrete samples with 1 cm, 10 cm and 100 cm length required very different chloride concentrations to initiate corrosion. No corrosion was found at more than 2.4% chloride by weight of cement for the 1 cm sample. The 10 cm sample started corroding at an average of 1.5% chloride by weight of cement. The 100 cm sample started corroding at an average of 0.9% chloride by weight of cement.

This has profound implications in that it means that the steel reinforcement layout has a large effect on the results obtained both in experimental procedures and also, more importantly, in real structures. Also, large real structures are likely to perform substantially worse than may be predicted by laboratory testing with small samples. The reason for this behaviour was conjectured (Angst and Elsener, 2017) to be that the bigger the area of the steel reinforcement the more likely there was to be inhomogeneities on the surface of the steel where this anodic reaction can prosper.

Whatever the reason for the results, this experiment shows that the presence and layout of steel in the concrete also has a dramatic effect on the corrosion process and this should be considered highly important. Most of the durability experiments that have been undertaken to date have neglected this fact.

2.5 IONIC MOVEMENT IN CONCRETE

Modern structures in aggressive environments are increasingly being engineered to achieve a life expectancy based on the impenetrability of the concrete to chloride ions. These predictions are commonly based on Fick's laws of diffusion even though it is known that this is an oversimplification of the transport situation in concrete. It is known that advection (capillary suction) and ionic migration can occur in concrete and its widespread use is probably linked to its simplicity and ability within certain situations to predict reality with adequate precision. Fick's law is a concentration diffusion mechanism and to use it directly assumes that transport is occurring in concrete in exactly the same way as in an aqueous medium. This is possibly a reasonable assumption for a simple concrete sample but it is not so reasonable when there are electrical potential differences in the structure, such as between two steel bars in a reinforced concrete laboratory sample and the huge number of bars at different depths and orientations in a real structure. These differences in potential are routinely seen when electrode (half-cell) potential mapping the surface of a reinforced concrete structure. This requires using an electrode with a known stable potential in a grid over the concrete surface to measure the varying potential of the steel reinforcement directly under this electrode. These potential differences create an electrical field that promotes the movement of ions by electro-osmotic flow from the anode to the cathode or vice versa depending on their charge.

It has been known for more than two hundred years (Reuss, 1809) that water could be made to percolate through porous clay diaphragms through the application of an electric field. The clay in this case behaves in the same way as concrete in that the particles acquire a surface charge when in contact with an electrolyte. Water is not electrically neutral and is a polar molecule. This means there is a small net negative charge near the oxygen atom and partial positive charges near the hydrogen atoms. So when there is a net charge in the electrolyte the water molecules align themselves in a polarised way on the solid particles which in this case will be the pore walls of the concrete. The immobile surface charge in turn attracts a cloud of free ions of the opposite sign creating a thin layer of mobile charges next to it. This electric double layer (EDL) is typically 1–10 nm and is commonly called the Debye layer (Figure 2.6). In the presence of an external electrical field the Debye layer acquires a momentum which is then transmitted to adjacent layers of the pore solution through the effects of viscosity. This is the process of electro-osmosis.

Evidence for this mechanism in reinforced concrete is provided by the fact that it has been known for many years that the application of an external direct current can dramatically alter the rate that ions, such as chloride, can move through concrete and this technique is used to make rapid assessments of the likely permeability of concrete samples. In a test such as the one commonly referred to as the 'Coulomb test' (ASTM C1202, 1997), the thickness of the sample is 50 mm and the test is run for six hours at 60 volts. The amount of chloride moved is the equivalent to that which would be expected in 50 years so that constitutes a rate increase of 73,000 times relative to straight diffusion. The test layout is shown in Figure 2.7.

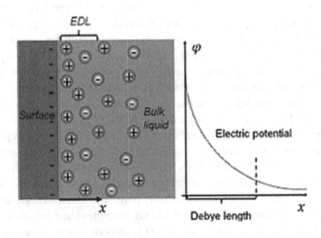

FIGURE 2.6   Electrical double layer.

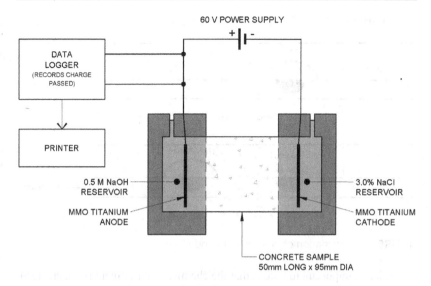

FIGURE 2.7   Drawing of Coulomb test layout.

In the last few decades electrochemical treatments such as electrochemical chloride removal and impressed current cathodic protection (ICCP) of reinforced concrete structures have become increasingly commonplace and are now being routinely installed on many projects around the world. In both of these processes there are significant flows of ions at far higher rates than would be predicted by diffusion calculations.

2.6 INCIPIENT ANODES – ADDITIONAL EVIDENCE OF ELECTROCHEMICAL EFFECTS

It has been commonly found in repairing chloride-damaged structures by patch repair with fresh mortar or concrete (without the use of rebar coatings) that the steel in nearby contaminated concrete then suffers from accelerated corrosion (Broomfield, 2007). The reason for this corrosion occurring in these particular types of patch repairs is that the previously corroding anodic areas become cathodic as they are covered with the high alkalinity mortar while the previously cathodic areas in the surrounding structure have then become anodic (Figure 2.8).

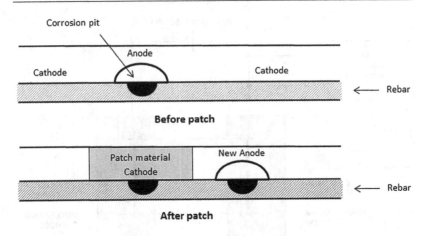

FIGURE 2.8 Formation of an incipient anode.

What is important to note is that the chemical environment is not changed at the previously cathodic location where the corrosion is now occurring. Although the immediate physical environment is unchanged, the only thing that has changed is that the potential difference is altered. Thus, this is the determining factor which is deciding whether corrosion occurs or not. With this potential difference there will be electro-osmotic movement or electromigration and this will after a short time alter the composition of the concrete at the steel interface. The other factor that the incipient anode effect demonstrates is that a potential difference change of less than half a volt (the difference between a typical potential value of passive steel in clean concrete and a typical potential value of active steel in chloride contaminated concrete) is enough to trigger this change from no corrosion to corrosion. This corrosion can be simply avoided by coating (insulating) the rebars before placing the mortar.

2.7 ADVECTION

This is the transport of material by bulk motion. For reinforced concrete the most important part of this is the transport of chloride ions by water movement through the concrete though many other dissolved ions will be moved also. This advection is caused by capillary suction which is the capillary pressure function of unsaturated porous material. When the concrete is dry there is a significant suction pressure built up. This can be easily demonstrated by pouring water onto dry concrete where it is immediately adsorbed.

For brick-built houses the salts in the groundwater can be carried up to a metre above ground level and then damage the brickwork when they precipitate as the water is evaporated. To prevent this advection, all houses use a damp-proof course (DPC). Stone-built structures also suffer with this problem. The purpose of this DPC is to prevent this cryptoflorescence causing damage. Due to the importance of this degradation mechanism the topic has been widely studied (Hall and Hoff, 2012) with findings that on London structures without a DPC damage occurred at 0.6–1.7 m above ground level while in Cairo the worst damage was 1.2–2.8 m above ground level.

The same mechanism also exists in concrete. If the wet part of the concrete structure is in potable water then there is not a significant problem as the dissolved salts from the groundwater precipitate into the concrete matrix.

If the wet part of the concrete is in saline water then there can be significant structural problems caused as the chloride depassivates the steel reinforcement. The big problems with this rapid movement of saline water come when there is a constant supply of salty water on one side and the concrete is dry and hot on the other side of the structure. For example, this situation applies to structures such as swimming pools which have sodium chloride and hypochlorites added to the water and have service tunnels around the walls of the pools which are hot and dry due to the machinery.

Another category of structures at risk are underground car parks or similar, particularly in the Middle East where there is saline groundwater and high ambient temperatures. The thinner the structure the quicker is seen evidence of chloride-induced damage of the steel reinforcement on the inside layers of reinforcement. This process can be rapid with severe corrosion recorded less than ten years after construction.

2.8 CHLORIDE PENETRATION – ADDITIONAL EVIDENCE OF ADVECTION

One of the most common ways of determining the penetration of chloride is to take dust samples of the concrete and then chemically analyse this for chloride levels at incremental depths. This is a quick and reliable test which is cheap and is performed on almost every survey of a reinforced concrete structure, so has been undertaken millions of times. In most samples of structures there is a significant deviation from the curve which would be expected from a purely diffusion-based ingress (Figure 2.9). In reality the chloride level measured is initially low then peaks, then slowly reduces in a manner that more closely follows a diffusion curve. In most cases this discrepancy has been ignored or

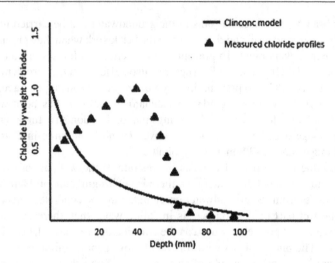

FIGURE 2.9   Actual chloride penetration curve against that predicted by diffusion calculation after 30 years' exposure on a bridge.

attributed to rain washing the surface of the concrete. This ignores that these same profiles are found, for example, on the underside of marine jetties which are protected from rainwater.

This has been acknowledged by Luping et al. (2012) who suggest that the first 15 mm of the concrete is considered a 'convection zone'. More accurately this movement should be called an 'advection zone'. Later in Luping et al.'s book, a mathematical model is compiled which considers an electrical field created by pore chemistry in this convection zone moving the chloride cation which is being measured. Perhaps this model is overcomplicated in that what is simply being recorded is the result of the outer layer of concrete having a different moisture level and chemistry to the surface layer which promotes capillary suction (advection) from the surface inwards.

2.9 MECHANICAL MOVEMENT

Most reinforced concrete structures are exposed to a changing environment. Some of this change can be caused by the weather and some can be movement caused by mechanical loading of the structure. When a structure is shaking and the environment contains chloride it has been found that there is a significant level of chloride injection by the opening and closing of cracks. This process can be so significant that corrosion is caused in a fraction of the time calculated.

2.10 STUDIES OF CORROSION ON STRUCTURES

A seasoned investigator (Gjorv, 2014) looked at many marine structures in Norway including harbour wharves. He found that different designs of reinforced concrete wharf in exposed marine conditions had massively different durability outcomes despite these being built at the same time with the same concrete, steel reinforcement and cover depths. He concluded that the reinforced concrete structures exposed to a chloride-containing environment would develop a complex system of galvanic cell activities along the embedded steel. In such a system the more exposed parts of the structure such as the deck beams would always absorb more chlorides and hence have anodic areas, while the less exposed areas such as the slab sections in between would act as cathodes. He found that wharves constructed without these beams were massively more durable with no corrosion on the bottom surface after 70 years.

Gjorv looked at a recent cruise terminal marine wharf beams, with an example given in Figure 2.10, and found that the steel in the reinforced deck beams was corroding within eight years after construction despite the concrete having a water/binder ratio of 0.45, a 28-day compressive strength of 45 MPa, silica fume addition and an average cover depth of 50 mm. No sign of this corrosion was given by electrode (half-cell) potential mapping, cracking or

FIGURE 2.10   Premature degradation of exposed marine piers.

rust-staining of the concrete. From conventional wisdom on the performance of steel in reinforced concrete there should be no corrosion after such a short time.

Another seasoned investigator (Melchers, 2017) has also looked at many reinforced concrete structures and came to the following conclusions. Very high levels of chloride concentration on reinforcing bars does not necessarily equate to corrosion. High quality concretes onset of serious, damaging corrosion is considerably later in time than its initiation. The causative mechanisms for the commencement of serious reinforcement corrosion remain unclear but do not appear to be the same as those for corrosion initiation. Furthermore, very severe localised reinforcement corrosion can also occur without obvious external signs of corrosion such as rust-staining and concrete cracking or spalling.

What both these studies demonstrate is that there is a great unpredictability in the way that structures behave in ostensibly similar environments. This leads to a conclusion that it is the design of the reinforced concrete structure which has the most significant effect on its corrosion durability in an aggressive environment. So if a structure was designed to have a similar potential of all the steel in the reinforced concrete elements then the rate of corrosion could be very low even at high chloride levels.

The findings reported above imply that there is a significant and dominant electrochemical effect on the corrosion process in real reinforced concrete structures and the preoccupation with chloride levels is incorrect.

2.11 CONCLUSIONS

When steel reinforcement is added to the concrete it can fundamentally change the behaviour of the concrete in that ions are moved not just by concentration diffusion but also by electromigration and by advection. These additional movement mechanisms have many important implications for the study of durability and also should have an influence on the best design practice for structures.

One of the more surprising aspects which has become evident is the speed of movement of ions through concrete in the presence of an electric field; perhaps this is because when we look at concrete, we have the impression it is totally inert and this biases our thinking. The possibility of electro-migration or advection has not been considered by any of the international standards with which reinforced concrete structures are routinely designed to comply.

The mechanism of corrosion of steel in concrete is complex and changeable with many factors having a significant effect on its occurrence and form.

This makes the widespread adoption of a single critical chloride level to denote corrosion/no corrosion scientifically invalid.

REFERENCES

Angst U and Elsener B (2017), The size effect in corrosion greatly influences the predicted lifespan of concrete infrastructures, *Science Advances*, 3, e1700751.

Angst U, Elsener B, Larsen C and Vennesland O (2011), Chloride induced reinforcement corrosion: Rate limiting step of early pitting corrosion, *Electrochemica Acta*, 56, 5877–5889.

Angst U et al. (2009), Critical chloride content in reinforced concrete – a review, *Cement and Concrete Research*, 39, 1122–1138.

ASTM1 C1202- 97 (1997), Standard Test Method for Electrical Indication of Concrete's Ability to Resist Chloride Ion Penetration.

Broomfield J (2007), *Corrosion of Steel in Concrete*, Second Edition, p. 120, Taylor and Francis, ISBN 0-415-33404-7.

Geiker A, Gehlen M et al. (2017), The steel concrete interface, *Materials and Structures*, 50, 143.

Gjorv, O (2014), *Durability Design of Concrete Structures in Severe Environments*, CRC Press, ISBN 13-978-1-4665-8729-8.

Green W, Collins F and Forsyth M (2017), "Up-to-date overview of aspects of steel reinforcement corrosion in concrete, (In) *Reinforced Concrete Corrosion, Protection, Repair and Durability*, (Eds) W K Green, F G Collins and M A Forsyth, Australasian Corrosion Association Inc, Melbourne, ISBN 978-0-646-97456-9.

Hall C and Hoff W (2012), *Water Transport in Brick, Stone and Concrete*, Spon Press, ISBN 978-0-415-56467-0.

Luping T, Nilsson L and Basheer P (2012), *Resistance of Concrete to Chloride Ingress*, p. 44, Spon Press, ISBN 978-0-415-48614-9.

Melchers R E (2017), "Modelling durability of reinforced concrete structures, (In) *Reinforced Concrete Corrosion, Protection, Repair and Durability*, (Eds) W K Green, F G Collins and M A Forsyth, Australasian Corrosion Association Inc, Melbourne, ISBN 978-0-646-97456-9.

Reuss F (1809), *Mem Soc Imperiale Naturilists de Moscow*, Second Edition, 327 F, Moscow University Press, Moscow, Russia.

Monitoring Corrosion and Why Most of the Current NDT Techniques Are Flawed

3

3.1 INTRODUCTION

There are several causes of degradation of reinforced concrete structures but this chapter will consider only the corrosion of the steel reinforcement. The primary cause of this is chloride contamination on the surface of the steel. Another degradation mechanism is carbonation of the concrete. This lowers the pH which destabilises the passivity of the oxide barrier on the steel and allows the corrosion process to occur at a significant rate.

The questions which are required to be answered for effective monitoring are: Is the steel corroding at an appreciable rate? and How much steel is left? With this information you may be able to make some form of estimate on the residual life of the structure. This estimation is not as simple as it may first appear, as certain structures can cope with massive amounts of corrosion while others only require a very small amount to require replacement or some other form of remediation.

The simplest way to directly answer the above questions are to knock off the concrete and visually examine the steel reinforcement and then measure directly the loss of section, pit depths and geographical layout of the corrosion problem. This will give most of the answers required above. One thing it will not be able to discern is when the corrosion started and thus the actual present corrosion rate cannot be determined. It has been practically found that corrosion initiation and propagation are separate events (Melchers, 2017) which makes the actual corrosion rate even more difficult to determine. Of course, this corrosion rate is also likely to vary in accordance with the rebar temperature and the moisture content of its immediate environment so in reality only a crude assessment can be made.

Unfortunately to remove a substantial proportion of the concrete cover over a structure's entire area is not practical in most investigations as the amount of damage caused will be large and very costly to repair. Normally this means that this destructive evaluation technique can only be used at selected locations or where sufficient corrosion has occurred to crack the concrete so it was going to be replaced in any event. Various other indirect techniques have been developed and they can be split into three broad groups which will be discussed further below, but all suffer from three significant problems:

1) Concrete is a heterogeneous structure with discrete lumps of aggregate, voids and differing chloride, alkali, moisture and oxygen levels. These will vary between the surface of the concrete to the rebar level where the corrosion is actually occurring.
2) The corrosion process of steel in concrete is a diverse, complex series of reactions with different intermediate corrosion products and cannot be categorised into a single process.
3) The actual steel reinforcement alignments and relative locations make a large difference to the likelihood and corrosion rate of the steel and contribute to a significant uncertainty with surface-based testing.

So the reality is that all the non-destructive testing (NDT) which is presently available can only, at best, give some indication of whether corrosion is occurring. Despite some techniques in a laboratory being able to give a rate of corrosion, getting a reliable result in the field is hoping for far too much, despite the huge commercial desire to accurately record this kinetic process.

3.2 CHANGES IN THE CONCRETE

3.2.1 Visual Cracking

The simplest and cheapest form of corrosion monitoring is to look at the concrete and identify any cracking on the surface of the structure – refer to Figure 3.1 as an example. Generally, this will be in the most vulnerable areas of the structure where the underlying reinforcement has become anodic to the rest of the structure. For example, this may be where salty water has been pooling. Another common area of corrosion is where there is restraint, shrinkage cracking or mechanical loading which allows a direct path to the steel reinforcement.

Sometimes this visual crack survey is enough monitoring, as if there are no other signs of corrosion-induced degradation then the owner will consider that the structure is fit for purpose. One of the weaknesses of a visual survey is that significant corrosion may not yet have built sufficient corrosion product to crack the concrete or that the corrosion product is not expansive and so despite extensive corrosion no surface degradation is seen.

3.2.2 Staining

Often associated with concrete that is either micro- or macro-cracked is a significant bloom of dark red iron oxide staining (Figure 3.1) which appears on

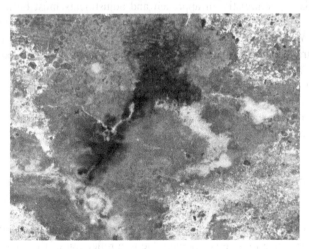

FIGURE 3.1   Reinforced concrete structure showing cracking and staining.

the surface of the concrete. This surface stain is an insoluble final oxidation product which has been transported as a soluble intermediate corrosion product through a crack or other defect to the surface where the higher oxygen availability causes the precipitation.

This visual test is simple, cheap and fast and is typically undertaken at the same time as a crack survey. The problems with this test are that the absence of stain on the surface does not preclude corrosion of the reinforcement and that you can often get staining on the surface which comes from tie wires or similar which are not at the rebar level and thus are not of structural concern.

3.2.3 Chloride-Sampling

This technique is a destructive sampling technique but the damage is minor. Normally a drilled sample is taken at incremental depths. The powdered concrete collected is then sent off to a specialised laboratory. At the lab the sample is reacted with acid and then titrated to get a total acid-soluble chloride concentration which is then sent back to the client.

There is a significant level of scatter between laboratories for this test. There are other techniques which can be used directly in the field to measure the sample chloride content, but these have become less popular in the last decade.

The chloride level measured is then directly correlated with a corrosion risk, i.e. at a certain percentage level (for example 0.4% by weight cement) corrosion will be initiated and will have a significant corrosion rate. It is known that this is too simplistic an approach and adjustments must be made. For example, an assessment of bound chlorides has been incorporated in some analyses. What is not adequately reflected is the huge difference to the chloride required for initiation and propagation, with this being discussed in more detail in Chapter 2.

What this means with chloride testing is that if you have no chloride contamination in concrete you will not have corrosion caused by this ion. Beyond this you can only surmise that as the chloride level rises there is more likely to be corrosion, but this is not universally so.

3.2.4 Hammer, Chain Testing or Ultrasonics

These are echo tests which aim to discover delamination or cracking. The most used is the hammer test where the surface is lightly knocked and an audible difference in tone is detected due to a reflection of the soundwave at the discontinuity.

The chain being dragged over the surface is a way of increasing the surface area rate being tested though it diminishes the acuity.

The impact echo is an electronic version of the hammer test that sends a pulsed soundwave into the concrete and records the echo time which is returned in the event of coming to a crack or significant voidage.

These tests all work quite well, though practically, Author 1 has always found the chain technique difficult to administer. The interpretation of all data is problematic as you are never sure of the cause of delamination. For example, delamination could be a cold joint or active corrosion and you are unsure of the actual situation without concrete removal or other information.

3.2.5 Carbonation Testing

This is a destructive technique where a chunk of cover is broken off and phenolphthalein pH indicator solution is sprayed onto the concrete. This is a quick and accurate test. Again, the interpretation of this data against the likelihood of active corrosion is the difficulty. This is because even if the low alkalinity layer does penetrate to the reinforcement level you are not sure that this will initiate corrosion at a significant rate.

3.2.6 Resistivity Testing

A probe system is placed on the concrete surface, which is typically arranged as a Wenner array with four pins, as shown in Figure 3.2, at several locations to allow a resistance map to be composed. There is also a simpler and cruder version with two pins. In the four-probe arrangement, current is passed between the outer two pins and the inner two pins measure the voltage drop. This resistance is a function of the probe spacings which is then turned into a resistivity. This is a dimensionless value normally reported for concrete as ohm.cm.

One of the most important factors in the resistivity of concrete is its water content. Concrete which is bone-dry at the level of the reinforcement may have a low corrosion rate even if there is chloride and oxygen. This is because the corrosion process is an electrolytic reaction. When the concrete is completely water-saturated there may be significant corrosion but this depends on several contributing factors.

This resistance test is simple and cheap to undertake but it does have some significant problems. One of the most relevant is that you are reading from a high resistance surface but what you are interested in is the condition of the concrete at, or near to, the steel interface and this may not be adequately

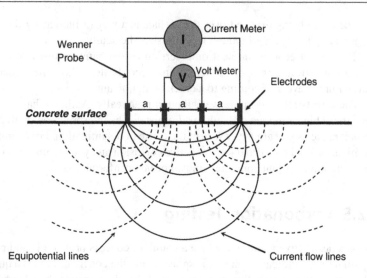

FIGURE 3.2   Four pin resistance measurement schematic.

represented. There are techniques available (Peabody, 2001) such as the Barnes method whereby varying the pin spacing to identify the concrete resistance at the depth of cover is possible, but they are not commonly used in the concrete-testing industry. This is because it is slow, as you have to change the pin spacings, and requires some mathematics.

Another big problem with the resistance testing method is that there are significant amounts of rebar in various orientations under the test area. This presence can significantly skew the results.

Despite several efforts there is no realistic pass/fail criteria for concrete resistivity measurements. What this means is that the resistivity at best is an indicative measure of the likelihood of corrosion which is rarely undertaken on its own.

3.3 CHANGES IN THE STEEL REINFORCEMENT

3.3.1 Half-Cell (Electrode) Potential

This is the most common form of NDT for steel reinforced concrete structures and has several significant advantages in that it is quick, cheap to perform and there are defined, though incorrect, pass/fail criteria. A reference electrode which is now typically silver/silver chloride (copper sulphate electrodes get poisoned by chloride ions) is moved around the surface and this fixed electrical potential allows the potential of the steel reinforcement nearest to the reference electrode to be measured against an electrical connection to the reinforcement.

The half-cell (electrode) potential which is being recorded is a combination of the anode corrosion potential and the cathode reduction potential of the steel reinforcement and typically the potentials will be a combination of the outer layer of steel reinforcement to an area of about 0.5 m², though the electrical resistivity of the concrete has a significant effect on the weighted radius that is being recorded.

The effect of surface moisture content on the measurement circuit can be seen in Figure 3.3. This shows a difference of 70 mV between dry and wetted for 15 minutes which is why wet sponge caps should be used on the surface of reference electrodes and 10 mV is a reasonable maximum resolution.

In general, steel which has a relatively positive (less negative) potential is unlikely to be corroding whereas steel with a negative potential may be corroding. The potential map does allow areas of the structure which are anodic to be identified and these relative potential maps are easy to construct, interpret and present.

There are conventions on how this testing is undertaken (ASTM, 2015) and the results classified. This interpretation is based on data obtained from North American bridge decks of potential recorded against a removal of the concrete and direct visual evaluation of the state of the rebar. Directly translating this interpretation to other structures, concrete mixes and environments without verification should be avoided (Figure 3.4).

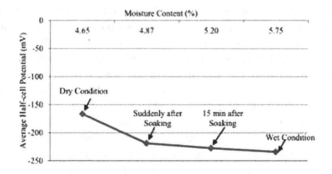

FIGURE 3.3 The effect of surface layer concrete resistance on a potential reading.

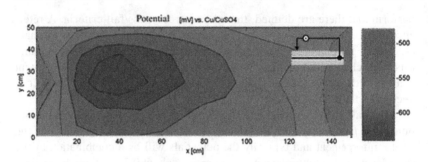

FIGURE 3.4 Typical half-cell (electrode) potential map.

3.3.2 Kelvin Probe Potential

The Kelvin probe is a non-contact vibrating capacitor device which measures the potential difference between a vibrating electrode, which is typically of stainless steel, and the reinforcement below the probe. In principle it has several advantages over the half-cell potential method outlined above as it removes the resistance errors in the measuring circuit through the concrete and it can also be used without a direct connection to the steel reinforcement. The probe requires a vibrator, which is typically a voice coil acting on the electrode, and a measuring circuit which is comprised of a zero resistance ammeter (ZRA) which can be adjusted using software. At this stage in its development (Sagues and Walsh, 2012) it appears to be critical that the concrete cover depth from the sensor to the reinforcing steel is constant.

3.3.3 Current Measurement

In principle it is quite easy to determine the amount of corrosion that is occurring as the metal corrosion gives off electrons which can be measured as an electronic current. The amount of current produced is directly proportional to the amount of the corrosion process so a reaction rate can be established. In practice to measure this in a structure and not a reinforced concrete test specimen purposely constructed for current measurement is difficult. This is because you cannot place new sensors into existing structures, as they almost certainly will not have the same steel surface conditions. Also, you could well be missing microcell corrosion where the anode and cathode are very close. In practice all you may be able to discern is macrocell corrosion.

The only practical thing you can do is isolate a single rebar or section from the structure where you think it is more likely to have active corrosion and measure the current flow from this piece of metal onto the rest of the steel reinforcement through a ZRA. This has only been tried on research projects (Chess, 1993) and even then, not that successfully. So, despite its promise of a direct real time reading of the actual corrosion rate being very appealing, it is not commonly used presently.

3.3.4 Polarisation Resistance

This is based on the finding that at a potential close to the freely corroding potential of a metal in an aqueous solution (typically +/− 20 mV), the polarisation slope is relatively linear and the gradient of this tangent is related to the corrosion rate. This is true for certain metals in certain solutions in a laboratory. However, even in the laboratory situation there are many metals and solutions which deviate from this tangential behaviour.

The situation for the cathodic and anodic reactions for steel in concrete where multi-stage reactions are combined with the impedance behaviour (you cannot correctly categorise concrete as having a solution resistance) of the concrete also mean that there is very little linearity and consequently this test should not be called linear polarisation resistance.

Commercially available polarisation resistance equipment normally has a guard ring for inserting the current onto the reinforcing steel commonly called a counter electrode; a reference electrode in the middle of this ring is shown in Figure 3.5 with the working electrode being the steel reinforcement.

The potential is typically stepped from −10 mV to +10 mV and the amount of current required to do this is recorded and processed and then a corrosion current derived in accordance with the Stern Geary equation. The galvanostatic

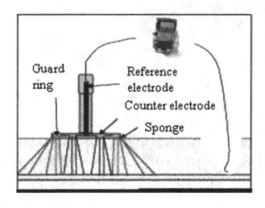

FIGURE 3.5   Typical polarisation resistance test setup.

method where the current is stepped and the changes in potential are recorded is also in use.

There are several obvious drawbacks to this test, but it is fairly quick and in practice affords some idea of the likely corrosion rate. However, the data should be treated with caution.

3.3.5 X-Ray

X-ray imaging always requires access to both sides of the structure. Typically, hard X-rays or Gamma rays are generated by an X-ray tube and introduced into the reinforced concrete element which are typically a suspended slab, wall or floor. The detector in the past was a chemical film but now it is commonly a digital device which allows digital imaging. The Xrays fall off in intensity as an inverse square of distance so a reasonable maximum separation between the source and detector is about 350 mm.

Concrete and corrosion products are significantly less opaque to the X-rays than steel so the rebar shows clearly as given in Figure 3.6. Using the image to detect a change in the profile of the rebar means that a resolution of at least about 1 mm is the best that can be achieved.

The advantages of this technique are that it allows direct verification of the bar's location and state. Its disadvantages are that it can only be used on certain structural arrangements, it has health implications for the operatives and has to be used in compliance with International Atomic Energy Authority regulations. This testing is expensive and not widely available outside the USA.

FIGURE 3.6   Picture of rebar arrangement and diameters captured by X-ray imaging.

3.3.6 Radar

Various radar trials have been undertaken but the resolution of line of sight radar is not that good and it has not proven as effective as X-rays due to its longer wavelength in the electromagnetic spectrum.

3.4 CHANGES IN THE STEEL TO CONCRETE INTERFACE

3.4.1 Heating Insulation

In this technique (Kobayashi, 2011), heat is applied to the rebar by induction heating through a coil and the behaviour of the rebar is studied by infrared thermography. The principle for this testing is that the thermal conductivity of corrosion product is less than that of the steel metal by two orders of magnitude and less than the concrete by one order of magnitude. In laboratory testing, differences could be measured in the heat distribution between samples with differing levels of corrosion product. This technique does not seem to have made it to widespread acceptance but recently a Greek university has also been publishing about this technique. The difficulties in applying this to a practical test regime are obvious in that it requires a significant power source to heat the steel reinforcement to a level that the thermographic cameras can detect, which gives rise to the possibility of differential thermal cracking and micro-cracking.

3.4.2 Electrochemical Impedance Spectroscopy (EIS)

EIS is primarily used in the corrosion industry for the accelerated evaluation of protective coatings on steel where it is evaluating changes at the steel to coating interface. These scans are typically 12 to 24 hours in length. In this technique a small voltage of between 5 and 50 mV is applied from an electrode on the surface at a varying frequency of between 0.001 and 100,000 Hz in an alternating current wave. This sweep is then compared electronically to the profile of steel which has been demonstrated to be corroding at other locations.

How this works is that most materials do not behave as a perfect resistor and at different frequencies their resistance varies. This variance in imped-ance is dependent on the composition and organisation of the layers of atoms directly on the steel surface. The take up of this technique is low because it is slow and expensive and as yet it has not been properly researched for the field testing of reinforced concrete.

Electrochemical frequency modulation (EFM) is an evolution of the above technique where two different alternating frequencies are introduced onto the reinforcement. Its principle advantage is that the Tafel slopes can be directly estimated with this reading.

Either of these techniques may in the future offer some practical advantages but the commercial systems offered to date for field reinforced concrete application have been poor.

3.4.3 Electrochemical Acoustic Noise

When oxide formation occurs it can have a large size difference to the original steel and an acoustic event will be generated. In the laboratory limited success has been had with this testing where microphones are strapped to the concrete and various frequencies monitored. This technique has not made it commercially, which is in contrast to its widespread use for the corrosion of the steel wires which form hanger cables on large bridges. Due to the limited area it can cover even with an array of microphones and its high cost it does not appear to be a strong contender for future development.

3.4.4 Connectionless Electrical Pulse Response Analysis

A probe consisting of 4 pins in the Wenner arrangement is applied to the concrete surface. The outer two probes have an AC current applied where the frequency is changed. The voltage (Giatec, 2019) between the two inner probes is measured as the frequency of the AC is changed by applying a narrow current pulse. It is claimed that the voltage response at different frequencies of a corroding bar are different to that of a non-corroding bar and this is picked up by the two inner probes. No discussion on the rebar's alignment, depth of cover or effect of the impedance of concrete is made. In EIS a complete spectrum is required, where here a pulse is deemed to suffice.

3.5 SUMMARY

The corrosion of steel in concrete has caused severe degradation of infrastructure and other structures around the world. The present most effective method of estimating the damage is selective removal of the concrete cover with structure-wide visual survey, some half-cell (electrode) potential mapping and perhaps some chloride samples.

There has been a trend in recent years to put huge resource into a multi-faceted corrosion survey which has been of little use in moving the remediation of the structure forward. Generally, it is apparent from a very limited survey, for example, of a wharf that has been in a marine environment for 30 years, that the corrosion apparent has been caused by chloride ingress, and also that the worst affected areas are the piles and supporting beams and that where the concrete has spalled 10% of the diameter of the bars has been lost. This information will be sufficient to be part of the remediation process described in Chapter 6.

REFERENCES

ASTM C876-15 (2015), Standard Test Method for Corrosion Potentials of Uncoated Reinforcing Steel in Concrete.

Chess P, November (1993), Cathodic protection of pre-corroded reinforced concrete, Spencer & Partners contractor report for UK Transport and Road Research Laboratory.

Giatec iCorr online website, March (2019), Commercial and Technical Advertising Document.

Kobayashi K, Nakamura R Rogugo K et al., September (2011), Concrete solutions, *4th International Congress on Concrete Repair*, Dresden, Germany.

Melchers R E (2017), Modelling durability of reinforced concrete structures, (In) *Reinforced Concrete Corrosion, Protection, Repair and Durability*, (Eds) W K Green, F G Collins and M A Forsyth, Australasian Corrosion Association Inc, Melbourne, ISBN 978-0-646-97456-9.

Peabody A (2001), *Control of Pipeline Corrosion*, Second Edition, p. 89, NACE Press, Houston, ISBN 1-57590-092-0.

Sagues A and Walsh M (2012), Kelvin probe for contactless potential measurements on concrete- properties and corrosion profiling application, *Corrosion Science*, 56, 26–35, Elsevier Press.

Solutions for New Structures

4

4.1 DESIGN LIFE IN DIFFERENT STRUCTURES AND COUNTRIES

In civil engineering it is common for the design codes to include a design life. These design lives have tended to get longer as the standards have evolved. Three examples of this for bridge structures are given here:

1) In the UK and Europe, the basis of structural design used (BS1990, 2002) gave a design life requirement for monumental building structures, bridges and other civil engineering structures of 100 years. This life requirement has since been increased to 120 years.
2) In the USA, there are bridge design specifications (AASHTO, 2012) which require a design life of 75 years and more recently a strategic highways initiative (SHRP2, 2013) which looked for a 100 year life and also provided an outline methodology of how to obtain this life.
3) The Norwegians started with a design life of 60 years and then went to 100 years (Gjorv, 2014) life requirement for their bridge structures.

In most countries around the globe the design life required is similar to the above standards. In the past, and particularly between 1950 and 1990, little thought was given to how to practically achieve this design life for reinforced concrete structures but recently the Americans (SHRP2, 2013) came out with a fault tree as shown in Figure 4.1.

The methodology of this fault tree dictates that, for example, the bearings have a life expectancy of 20 years, thus the design has to have a strategy to replace

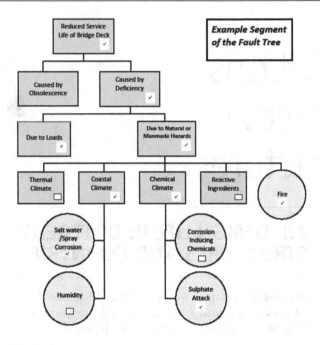

FIGURE 4.1   Fault tree.

them four times so that the structure can achieve 100 years. This strategy has very rarely been followed through with the main components of a reinforced concrete bridge, but is generally followed on steel bridges where a full coating repair and replacement strategy over several cycles to obtain the design life in accordance with this fault tree is normal. This anomalous behaviour can only be explained by the subconscious thought of the civil engineers who persist in the notion that standard reinforced concrete structures without additional measures can be durable to huge design lives in an aggressive environment. This is despite practical evidence that this is not the case with many prematurely decaying structures.

On smaller civil engineering projects these very long design lives are assumed to be achievable by simply being in compliance with the material codes. The specific exposure condition of the structure is rarely further considered. On a few large projects and those where recent durability issues have been highly visible, particularly where there is oil and gas industry involvement, then a more comprehensive review of the actual likely life expectancy and what can be done to attain the required figure has been made. This is because the oil and gas industry employ trained corrosion engineers, albeit they are generally specialists on steel structures and not steel in concrete.

In general, there has been a shockingly low duty of care in the civil engineering industry about the long-term durability despite the huge expense of

FIGURE 4.2 'Spaghetti Junction', Birmingham, UK.

replacement works which can be far more costly than the original construction. An example of this cost is 'Spaghetti Junction' in Birmingham, England (Figure 4.2). This is a complicated interchange of raised roads which was completed in 1972 at a total cost of £87.69 million for all 16 contracts. As part of this construction there are 21 km of elevated roadways with 1,200 spans, 1,100 reinforced concrete bents/crosshead beams/headstocks and 3,500 reinforced concrete columns.

This structure had a design life of 100 years. Despite this very long life expectancy, no thought was given to durability beyond complying with the relevant standards of the time for concrete strength and depth of cover. For the first 15 years after opening virtually no maintenance was undertaken apart from some seized steel to steel bearings being replaced, so effectively, it was abandoned. But when it was surveyed after only 15 years the results were really worrying with further bearing problems, expansion gap degeneration and deck waterproofing failures, to name some of the woes that needed urgent remediation. Perhaps the most significant concern was the active corrosion of the steel in the reinforced concrete bents and shear walls which support the spans. To combat this last problem a large-scale trial of cathodic protection was organised in 1987 and since its successful conclusion in 1991, more than 740 structures, typically bents, have been provided with impressed current cathodic protection (Christodoulou et al., 2014).

The deck ends were not surveyed due to difficulties in assessing this area despite their structural significance and the fact the chloride source of de-icing salt in rain water was coming through the deck at this location. An assessment of the damage overall was made and a priority list of repairs developed. A maintenance programme was instituted with rolling permanent maintenance with term contractors and individual contracts for major works.

While these have varied in size a typical individual contract is £5 million (CRL, 2019) for the repair and impressed current cathodic protection of 16 bents. Using this figure as typical of the 740 structures which have been cathodically protected from 1991 to 2014 the repair bill for these individual contracts and not the term contractors so far in today's money is £231 million. The structures are approaching half of their design lives and only a proportion of the structures has been life-enhanced. With hindsight it would have been vastly cheaper to optimise the durability of the structure at the initial design stage.

Unfortunately, the way civil engineers are trained means that the structural components are treated in great detail and the durability aspects are minimally addressed. So while it is vital that the structure is originally structurally sound, its ongoing durability may well define its life-cycle cost and this should be treated in the same rigorous manner as its loadings. It should, in fairness, be understood that the extremely long service life requirements referred to earlier in this chapter make it very difficult to properly corrosion engineer a reinforced concrete structure in a known aggressive environment.

4.2 MODELLING AND ITS LIMITATIONS

Some books have been written, for example, Luping et al. (2012), wholly about the resistance of concrete to chloride ingress and it has been tempting to engineers to take the diffusion rate values in validated models for a certain concrete mix type, plug-in the cover depth, decide what is an acceptable chloride contamination level at the outer layer of reinforcement and then compute that this life expectancy is beyond the design life and that's it – job done. This is still the most common way of approaching the design and even very experienced corrosion engineers (Gjorv, 2014) have followed this methodology.

Unfortunately, this approach is inherently flawed on many levels. The first deficiency is that it assumes concentration diffusion is the only ion movement mechanism whereas in a real structure (Chess, 2019) there can be other factors such as advection and electro-migration which are much quicker transport mechanisms. Incidentally, a recent finding is that the stress level on the hardened concrete has a significant effect in increasing the rate of movement of chloride ions.

The second problem with this approach is it requires a defined chloride level when significant corrosion will occur. This critical chloride level is actually very dependent on the reinforcement layout and its relative alignment and many other factors and can vary by a factor of 100 (Chess, 2019).

The final problem with the modelling process, which is typically completely ignored, is what level and type of corrosion is required to define the actual life of a structure. For example, an old column and beam bridge over a motorway may structurally be able to easily withstand a 30% loss of section of all its outer reinforcement on the columns and crossheads. So you should have many years after the start of significant levels of corrosion before you have to repair. However, the risk of concrete spalling onto passing transport vehicles as a result of this corrosion process might impel the owner to implement a full repair or replacement at a relatively early time as only a small amount of corrosion is enough to raise tensile stresses sufficient to cause spalling.

If this same structure is placed in a marine setting then apart from the aesthetics, no repair or replacement is required for many years beyond this point as the spalled concrete should harmlessly drop into the water.

At the other end of the risk spectrum are structures with pre- or post-tensioned steel tendons directly introduced into the concrete or placed in ducts. In these structures only very small amounts of corrosion to these tendons can be tolerated and it only has to occur in an isolated area for structural collapse to be a distinct possibility.

4.3 EVALUATING SIMILAR STRUCTURES IN SIMILAR ENVIRONMENTS

Presently this is the most effective method to use but it does have significant limitations. These are:

1) The structure type and specification can be too new for significant degradation to be observed or measured in comparable circumstances.

2) In certain structures, such as tunnels, with limited access it is almost impossible to see or measure the structural degradation that is occurring.

3) For political, financial or ignorance reasons the owners have no durability monitoring of an existing structure so cannot provide any information on its performance.
4) There is no budget for travel and analysis of similar structures when in the design phase of a new structure and no mechanism for incorporating this information into the final design.

Generally, the most helpful information is case histories of similar structures which have been repaired to help them achieve at least a proportion of their design life. For example (Gries et al., 2018) gives a description of the repairs to power station cooling towers in the USA on the coastline. These were built in 1976 and life-threatening levels of damage were being recorded by 2010. This gives an actual life of 34 years achieved before large-scale repair or replacement was required. The construction and design practices of these parabolic towers has not substantially changed over the intervening period so a similar life-span may be expected unless additional anti-corrosion measures are taken to gain the required enhancement in design life on new cooling towers.

In many cases a significant level of interpretation will be required to relate the data from an old structure to the new one. This will cause a level of uncertainty on the actual life of the new structure to be generated. For an example, it may typically be 30 years before there is evident structural damage such as cracking and staining on an exposed marine bridge pier. This time period before damage might reasonably be thought to occur should be significantly increased by the increase of cover, and the use of blended cement-based concrete on a new structure. However, what was actually found on a recent bridge pier with 90 MPa blended cement-based concrete and a 75 mm cover depth was that significant corrosion-induced damage of the steel had occurred within ten years of the bridge opening. The possible reason for this disparity from what was modelled was the varying wind loading of this element caused occasional tensile stress and the very low ductility of the concrete allowed significant cracking. These cracks opening and closing were actively pumping chloride through the cover depth to the steel reinforcement.

4.4 IMPROVING THE DURABILITY OF CONCRETE

The most common method used is to modify the pore morphology so the tortuosity increases and thus the diffusion rate decreases. Ways to do this which are commonly used are replacing the ordinary Portland cement (OPC) with pozzolans (e.g. slag, fly ash and silica fume) and reducing the water to binder ratio. This is all admirable as long as the workability is satisfactory so that an even coating surrounds the entirety of the reinforcement and the mechanical properties are kept intact. The property of concrete generally ignored which is of importance in durability is ductility. This is an important requirement as although the primary purpose of the steel reinforcement is to take the tensile loads, there can be tensile stresses, particularly in the concrete cover, which can cause limited deformation or cracking. Increasing the cement contents while helping increase strength can also promote shrinkage cracking, particularly where there is restraint in the structure. This has caused such large problems in American bridge decks they are now restricting the cement content of the concrete and using stainless steel reinforcement (Van Dyke et al., 2017) in Florida for example.

Another approach to increasing the durability is to add chemicals to the concrete which stop or reduce the corrosion rate in the future. Mostly these have been inhibitor chemicals with the most commercially successful being calcium nitrite. Introducing the chemical evenly into the mix is likely to prevent the problem found with inhibitors applied from the outside of the hardened concrete which means that the concentration on the rebar surface is always uneven. Another chemical which has been commercially offered is calcium nitrate. In the short term, the presence of these chemicals can significantly increase the chloride level which is required (Bertolini et al., 2013) to initiate corrosion, but once corrosion has initiated had no effect on the rate. About ten years' additional life expectancy seems appropriate by using an inhibitor or a pore blocker in the concrete with no other change.

4.5 IMPROVING THE DURABILITY OF REINFORCEMENT

There are a few ways of achieving this:

1) Coating the rebar

The two most common techniques have been an epoxy coating or a zinc coating but cladding with a more noble metal (typically stainless steel) has been used. Epoxy coating was being pushed hard by large chemical companies but its use outside the United States is now quite limited due to it having well-publicised problems of corrosion occurring below the coating. If this is the only change in a structure then getting a 10-year life increase before corrosion induced damage might be realistic though some real-world failures have shown that there was minimal life expectancy improvement (Van Dyke et al., 2017) over black steel.

Galvanised rebar is in quite common use around the world and a large volume of research and practical performance was summarised in a fairly recent book (Yeomans, 2004). In this the relatively poor performance of this material in laboratory trials was contrasted to the better performance in actual long-term installations. The reduction in performance when galvanised and black steel bars were included in the same circuit was also noted. From the performance of the bridge installations then, an additional 15 to 20 years life increase before corrosion-induced damage is realistic if all the rebar in the exposed areas is galvanised.

There has been a recent introduction in North America of an 'improved' galvanised product where the intermetallic zinc and iron layers are minimized despite the total zinc coating thickness being reduced. This is claimed to offer performance advantages despite having less coating.

Rebar cladded with more noble materials such as stainless steel was being manufactured in the USA in the past but concerns over bending the rebar and subsequent cracking of the outer layer were expressed and it is now understood to no longer be commonly used.

2) Changing the metal

In the past bronze was used successfully, though it is very rarely used presently due to its high price and limited availability. The most commonly used material is stainless steel which is now commonly in use in the United States, Europe and elsewhere.

In testing (Van Dyke et al., 2017) the use of a mix of stainless steel and black steel has not been found to have a deleterious effect on either material despite their different potentials. Traditionally, grade 304 which is the lowest grade austenitic alloy has been used and this seems to have performed well over extended periods on actual structures. In the laboratory, 316 grade, which is commonly known as marine stainless, and even better duplex have provided corrosion immunity in most concrete scenarios.

Duplex stainless steel has a significant advantage in that its mechanical properties is better than austenitic stainless steel grades and can be superior to high yield rebar. The improvement in durability from a corrosion viewpoint should be very large so life expectancy of more than 100 years can be postulated.

3) Changing the potential of the steel reinforcement

This is an active process where you cast in anodes to depress the potential of the rebar. Both impressed current and galvanic anodes are offered commercially for this purpose. When concrete is originally cast the resistance of the concrete is quite low but this will rapidly climb to a level which is higher than the driving potential for a galvanic anode to pass a requisite level of current for effective protection (Chess, 2019). The impressed current anode option is a much better technical solution but suffers from the significant drawback that it is required to be actively operated and maintained. If this is done and a decent job of installing the anode and wiring is made then a life expectancy improvement of more than 40 years can be anticipated.

4) Non-metallic reinforcement

Glass fibre-reinforced polymers in the shape of rebars have been increasingly common over the last 20 years and presently are being commercially installed in high risk areas. Other similar non-metallic materials such as basalt also appear to be promising.

The big unknown with these is the durability of the binder for the fibres which is typically a vinyl ester epoxy. It is known that epoxy on steel in chloride-contaminated concrete has deteriorated over a 20-year period and this degradation process is likely to be similar but how this will affect the functionality of the reinforcement is not yet clear (Boer et al., 2013).

4.6 IMPROVING DURABILITY OF THE STRUCTURE

The simplest way of undertaking this is to clad or coat the structure. This should prevent or more realistically reduce the penetration of the chloride to the surface of the concrete. This durability enhancement has an important advantage in that you can measure simply how well it is working by taking chloride samples from the concrete. Many types of coatings have been used, some more successfully than others. There are three broad categories of this as shown in Figure 4.3.

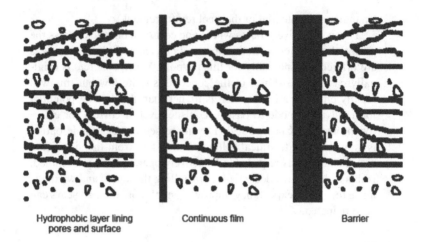

Hydrophobic layer lining   Continuous film   Barrier
pores and surface

FIGURE 4.3   The three categories of coatings.

4.7   HYDROPHOBIC OR THIN COATINGS

These are hydrophobic treatments that line capillary pores or reaction products that block the pores. The hydrophobic pores treatment is commonly a version of silane. These are liked by owners because they do not impose a maintenance schedule as you cannot see their deterioration. The problems come with their lack of penetration into wet concrete and relatively short time frame effectiveness. For example, Liu et al. (2005) had difficulty in getting any penetration into wet concrete but in drier material it was found that this treatment reduced the chloride ingress by up to 70% over a four-year period. In the USA this treatment is commonly reapplied at five-year intervals (Van Dyke et al., 2017), so as life enhancement treatment only applied during construction, its life expectancy enhancement is likely to be negligible.

4.8 CONTINUOUS FILM OR
THICK COATINGS

There is a whole host of coatings which are used for coating applications in this category such as epoxies, polyureas, polyurethanes and acrylics to name but a few. There have been many claims over their effectiveness but they must all remain adherent, not crack and be resistant to chloride ingress to be effective over their lifetime. It is difficult to give a lifetime for these coatings but using a standard for steel (DNV, 2017) there is an equation for coating breakdown (F_c):

$$F_c = a + b \cdot t$$

With a 250 μm thickness coating, a is 0.05 and b is 0.020 and t is time in years.

So taking a time of 20 years there will be a 45% coating breakdown.

Interpreting this is complicated, as in this standard the substrate is different and the environment is marine-exposed, but it gives some sort of indication of when a coating may reasonably be expected to become ineffective against chloride migration.

Using this data, a reasonable maximum recoating regime of less than 20 years will be required.

4.9 BARRIER

This is a category that includes tiling, sheet application and encasement within a metallic cladding.

Tiling is used in specific locations such as lining the inside of tunnels and for swimming pools. The reason for its use is probably more for decorative purposes then durability enhancement. Despite the glazed outer surface which you would expect to be impervious to chloride, substantial levels of chloride do penetrate and swimming pool walls are commonly repaired with impressed current cathodic protection to prevent ongoing chloride-induced corrosion of the reinforcement after about 30 years of service (Chess et al., 1999).

Certain categories of civil engineering have taken durability much more seriously than most of the industry where the structure is to be placed in a known hostile corrosion environment. A notable example of this is in the design of immersed tube tunnels (Lunniss and Baber, 2013) which have used steel sheets, high-density polyethylene (HDPE) sheets and bitumen sheets to protect the outside of immersed tube tunnel segments.

When they have used steel, they have also coated it and protected it further with galvanic anodes. With this arrangement a life expectancy of 20-plus years for the coating can be estimated, then an additional ten years for the galvanic cathodic protection. The next layer of protection is the wastage of the steel. The corrosion rate of this, after several years' exposure, is given as 0.05 mm per year (Baboian, 2002). This would give an additional 16 years, as the steel membrane is typically 8 mm thick in a marine exposed environment and significantly more in the more clement surroundings of the buried tunnel. So, realistically 30 years can be estimated before the steel is perforated. At this time the concrete cover will be brought into use.

The immersed tube tunnel outside is corrosion-engineered so no chloride penetration into the concrete will occur for the first 60 years. Due to the low oxygen conditions on the outside of the tunnel, corrosion normally occurs on the inner layers of the steel reinforcement. The immersed tube tunnel walls are exceptionally thick for buoyancy reasons and they may easily be able to last more than sixty years without degradation from the outside, so this is one of the very few structures which has been corrosion-engineered in an aggressive environment to realistically last more than 120 years.

The threat of corrosion on the inside of the tunnel is dependent on the location of the tunnel. In northern countries de-icing salt is carried into the tunnel in significant quantities by motor traffic and this problem is sometimes countered by liquid membranes but in some tunnels where the construction money ran short and there is no protection beyond concrete cover, significant corrosion problems are seen after 30 years.

Rail tunnels do not have the same levels of chloride problem and their difficulties can normally be attributed to AC or DC leakage from the traction systems combined with a much lower level of chloride which comes from leakage.

The other category where steel cladding is commonly used to protect the concrete is when a steel tube is used as a former for piling and left in place after casting.

To the best of the author's knowledge no reinforced concrete structure has been clad with stainless steel sheeting as a corrosion protection system. This material has only been used for aesthetic purposes. This type of cladding and other corrosion resistant materials such as titanium could provide a design life of more than 100 years.

4.10 CHANGING DESIGN

Recent publications (Chess, 2019) have demonstrated that the design of the structure will have a great influence on its longevity. It is well known that designing structures with ribs or beams which are the first to become chloride contaminated will then corrode preferentially (Figure 4.4).

So, in aggressive conditions an attempt should be made to remove all reinforced concrete elements which are going to become anodes to the bulk steel cathode in the protected part of the structure. As an example, see Figure 4.5 which shows fuelling or storage barges made from reinforced concrete.

FIGURE 4.4 Severe corrosion on underside of a marine jetty beam after 18 years.

FIGURE 4.5 Reinforced concrete barges built in 1940.

These barges have been in cyclical marine immersion conditions for almost 80 years, are riddled with chlorides and yet have very little corrosion and continue to be used in many ports around the UK. One of the reasons for this longevity is the similar potentials of all the reinforcement steel.

4.11 COMBINATION OF TECHNIQUES

Depending on the determination of the client to actually achieve the specified design life it will be worth visiting the possibility of using a combination of these life enhancement techniques. It is also possible to use different life enhancement measures at specific locations. For example, on the previously mentioned 'Spaghetti Junction' structure, the top layer of steel rebar on the bent, shear walls and deck ends could be substituted with stainless steel reinforcement. The design could be changed to promote runoff of the contaminated deck water and a thick mastic coating installed over all the surfaces of the bent and shear walls. While these amendments would significantly increase the initial cost and have a maintenance demand for the periodic replacement of the coating it will be dramatically cheaper than present maintenance costs.

REFERENCES

AASHTO LRFD Bridge Design Specifications July (2012), ISBN 978-1-56051-523-4 publication code LFRDUS-6.

Baboian R, (2002), *NACE Corrosion Engineer's Reference Book*, Third Edition, NACE Press, Houston, ISBN 1575901277.

Bertolini L, Elsener B, Pedeferri P, Redaelli E and Polder R (2013), *Corrosion of Steel in Concrete*, Second Edition, Wiley-VCH, Weinheim, ISBN 978-3-527-33146-8.

Boer P, Holliday L and Kang T (2013), Independent environmental effects on durability of fibre-reinforced polymer wraps in civil applications: A review, *Construction and Building Materials*, 48, pp 360–370.

BS EN1990 Eurocode –Basis of structural design (2002), British Standards Institute.

Chess P (2019), *Cathodic Protection for Reinforced Concrete Structures*, p. 7, CRC Press, ISBN 13-978-1-138-47727-8.

Chess P, Jakobsen D, Lawrence R and Godfrey J, (1999), *Cathodic Protection of Reinforced Concrete Swimming Pools*, BiNDT & CORR 99, Poole, Institute of Corrosion.

Christodoulou C, Sharifi A, Das S and Goodier, C. March (2014), Cathodic protection on the UK's midland links motorway viaducts, *Proceedings of the Institute of Civil Engineers –Bridge Engineering*, 167, 1.

CRL (2019), Concrete Repairs Limited, website 29/05/2019, news, CRL commence work on gravelly hill viaduct.

DNVGL-RP-B401 (June 2017), *Cathodic Protection Design*, Det Norske Veritas.

Gjorv O (2014), *Durability Design of Concrete Structures in Severe Environments*, Second Edition, CRC Press, Boca Raton, ISBN 978-1-4665-8729-8.

Gries M, Michols K and Lawlar J (2018), Corrosion 2018, NACE, Phoenix, USA, paper No 11005, Evaluation and repair of natural draft cooling towers.

Liu G, Stavem P and Gjorv O (2005), Effect of surface hydrophobation for protection of early age concrete against chloride penetration, Fourth International Conference on Water Repellent Treatment of Building Materials, pp. 93–104, (Eds) J Silfwerband, Aedifacatio publishers, Freiburg.

Luping T, Nilsson L and Basheer P (2012), *Resistance of Concrete to Chloride Ingress*, Spon press, Abingdon, ISBN 978-0-415-48614-9.

Lunniss R and Baber J (2013), *Immersed Tunnels*, p. 302, Taylor and Francis, Boca Raton, ISBN 13 9780415 459860.

SHRP2 April (2013), Report S2-R19A-RW-2, Design Guide for Bridges for Service Life, Federal Highways Administration.

Van Dyke C, Palle S and Wells D (September 2017), Long Term Corrosion Protection of Bridge Element Reinforcing Materials in Concrete, KTC-17-03/ SPR16-513-1F.

Yeomans S, (2004) *Laboratory and Field Performance of Galvanized Steel in Concrete*, Elselvier Science, 1st edition, ISBN 10978 008 044 5113.

Maximising Service Life with Minimal Capital Expenditure

5

5.1 PHASES IN THE LIFE OF A STRUCTURE

The service life of a structure can most certainly be maximised with minimal expenditure (operating and/or capital). The phases in the life of a reinforced concrete structure are summarised at Figure 5.1.

The maintenance phase in the operational service life of a structure can therefore be optimised given the wide range of protection, repair and durability options available including at minimal expenditure over the life-cycle of the structure.

5.2 OWNER REQUIREMENTS

This is the first thing that is critical to adopting a realistic design approach to maximising the operational service life of a structure. The common position is that the owner will state that they require the 100–120-year life in the design codes and would like you to properly design it to achieve this. The next part of this design approach will normally be down to the type of owner, their wealth and history. For example, with most government bridge owners when it is suggested that the piers are coated with a protective coating when built,

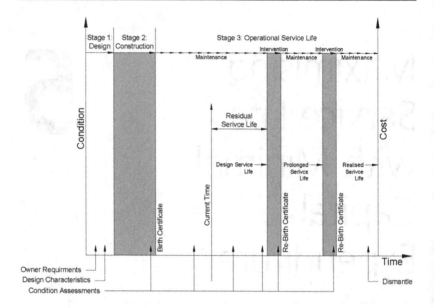

FIGURE 5.1   Phases in the life of a reinforced concrete structure. (From fib, 2012.)

they will come back with this raising the future maintenance requirement and thus being an unacceptable solution to enhancing durability. Perversely, they will allow a pore-blocking penetrant system, as while this is unlikely to have a long-term benefit you cannot see easily its degradation, so that does not lock them into a 20 year remove and replace coating commitment. In fact, government clients are normally unwilling to consider any life-enhancement procedures which may have an increased maintenance budget.

In contrast, agencies which are going to build, own, operate and transfer (BOOT) are very much more open to being innovative as in this case at the transfer period all signs of degradation are looked at to reduce the price the government has to pay.

If durability planning is adopted as an integral part of a project, whether it be a new build or an existing structure, then the asset owner can specify their requirements and also ensure that there is a continuous link in durability objectives from design through construction and into operation and maintenance. Asset owner key durability planning purpose and benefits are (Concrete Institute of Australia, 2014):

- Intended design life of asset is achieved with more confidence.
- Premature damage risk is reduced during construction and design life.

- Improved confidence in the technical capability of the design and contractor team via complementary durability consultant input.
- Maintenance requirements are better taken into account during design and construction, with durability-related matters having greater input into operation and maintenance manuals. A proactive or reactive maintenance strategy can be adopted with economic impact shown (Figure 5.2). The proactive strategy normally requires frequent expenditure resulting in an asset that stays in good condition throughout its operational service life whereas a reactive strategy is accompanied by major rehabilitation/repair costs when poor condition is reached.
- Whole-of-life asset cost is optimised. Early project formal durability input is more cost-effective than being introduced during detailed design or construction.

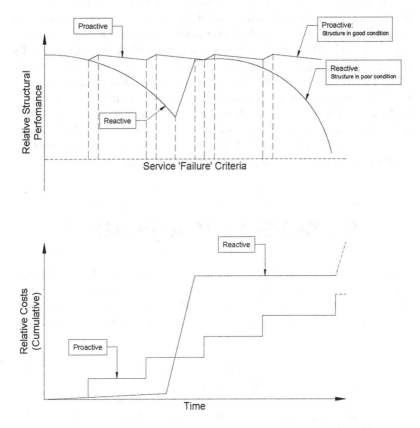

FIGURE 5.2   Relative costs for reactive and proactive maintenance. (From fib, 2008.)

- At the end of the structure or the end of design life, condition is acceptable to the asset owner and capable of prolonged service life with maintenance.

5.3 DURABILITY CONSULTANT

The durability consultant first has to understand whose side they are on and what is politically possible within the limits of the brief. Generally, on infrastructure and architectural works the look of the structure has the highest priority, and its durability and constructability are way down the list. The next part of the requirements for the specialist is to understand the construction process and its flexibility to be changed or improved. A critical part of this is where in the cycle of design and submission part of the process of the structure they are involved and how demanding the owner is for a costed and realistic durability approach for all the reinforced concrete components of the structure. Normally the most demanding are in the oil and gas industry who routinely life-cycle steel structures and have carried it over onto reinforced concrete structures. The least likely to undertake any durability analysis are local authorities on smaller bridges and other structures such as municipal swimming pools which have a dreadful durability record. In this latter case, an early part of the durability specialist's responsibility is to enlighten the client on the premature problems occurring in similar circumstances on similar structures.

5.4 DESIGNER REQUIREMENTS

Where durability planning is an integral part of a project, whether it be new construction or service life extension of an existing structure, the Concrete Institute of Australia Durability Planning Recommended Practice document (Concrete Institute of Australia, 2014) indicates that designer key durability planning purposes and benefits are:

- Durability input is provided that is not available from Standards and Codes, which lack durability certainty.
- Durability input is provided that is typically not available from Structural and Architectural Design staff, and is delivered in a complementary process.

- Construction materials selected by designers are more likely to achieve durability performance during design life.
- Designer durability-related mistakes or negligence are reduced.
- The design is more likely to achieve the asset owner's intended design life (as legally required in the designer's agreement with the asset owner) without premature durability-related damage during construction and the designer's project defects liability period.
- Design completed with first time appropriate durability input will provide cost benefit savings for the designer (e.g. less re-design staff time and litigation claims).

A durability process that could be adopted during the design phase of a new structure or the repair of an existing structure is summarised in flowchart form at Figure 5.3 (Concrete Institute of Australia, 2014).

5.5 CONTRACTOR REQUIREMENTS

The contractor has normally won the project on the lowest price and really wants to keep the costs to a bare minimum to maximise its profits. So they are caught in a dilemma: they want to satisfy the owner's requirements at the minimum cost and they do not want any deleterious effect on their schedule by enhanced durability measures.

Unless a contractor has been financially hit by durability issues from previous projects, they are not particularly interested in instituting a comprehensive durability assessment and they certainly want to kick the can down the road. So of some interest is a future maintenance regime which is reactive rather than proactive where the onus is put back onto the owner after a certain time period.

On the other hand, where durability planning is an integral part of a project, whether it be new construction or service life extension of an existing structure, the Concrete Institute of Australia Durability Planning Recommended Practice document (Concrete Institute of Australia, 2014) indicates that contractor key durability planning purposes and benefits are:

- Durability input provided that is not available from Standards and Codes. Project specifications require construction in accordance with these Standards and Codes, which lack durability certainty.
- Durability input provided that is typically not available from Structural and Architectural Design staff, and hence the contractor uses the durability consultant on request during construction.

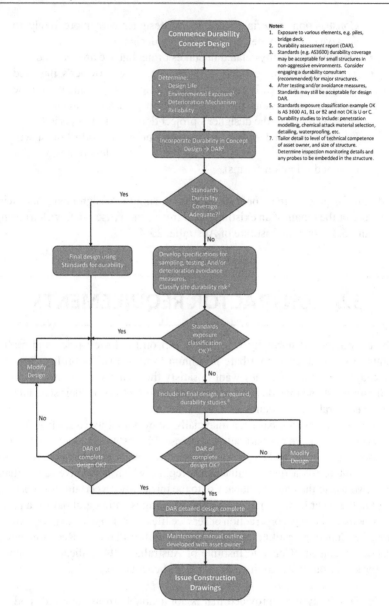

FIGURE 5.3 Durability process during design. (From Concrete Institute of Australia, 2014.)

- Construction materials used by the contractor are more likely to achieve durability performance during design life.
- Contractor durability-related mistakes or negligence are reduced.
- Construction is more likely to achieve the asset owner's intended design life (as legally required in the contractor's agreement with the asset owner) without premature durability-related damage during construction and the contractor's project defects liability period.
- Construction completed with first time appropriate durability input will provide cost benefit savings for the contractor (e.g. less construction defects, project completed without delays, less defects after construction and litigation claims).

A durability process that could be adopted during the construction phase of a new structure or the repair of an existing structure is summarised in flowchart form at Figure 5.4 (Concrete Institute of Australia, 2014).

5.6 OPERATOR/MAINTAINER REQUIREMENTS

Maintenance is an integral part of ensuring durability through the life-cycle of a structure, whether the structure is new or existing. An asset maintenance plan/asset management plan can be developed for a structure and can detail aspects such as (Concrete Institute of Australia, 2014):

- Details of asset management inspection audits including life-cycle program of tests to be undertaken, assessment criteria and actions to be taken related to the inspection and test results.
- Maintenance materials and methods to be adopted throughout the service life to achieve optimum whole-of-life cost.
- Maintenance and asset management planning based on materials deterioration expectations and costs to achieve acceptable asset owner performance predicted at appropriate intervals but not exceeding ten years.
- Identify repairs during maintenance audits to prevent further deterioration that may result in major problems.
- Minor repairs completed within required time schedule.

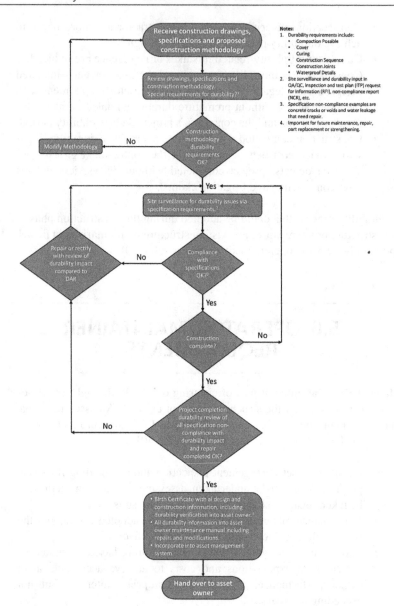

Notes:
1. Durability requirements include:
   - Compaction Possible
   - Cover
   - Curing
   - Construction Sequence
   - Construction Joints
   - Waterproof Details
2. Site surveillance and durability input in QA/QC, inspection and test plan (ITP) request for information (RFI), non-compliance report (NCR), etc.
3. Specification non-compliance examples are concrete cracks or voids and water leakage that need repair.
4. Important for future maintenance, repair, part replacement or strengthening.

FIGURE 5.4 Durability process during construction. (From Concrete Institute of Australia, 2014.)

- System for keeping records of deterioration and repairs throughout structure life, including repair materials used.

A durability process that could be adopted during the service life (operation and maintenance) phase of a new structure or the repair of an existing structure is summarised in flowchart form at Figure 5.5 (Concrete Institute of Australia, 2014).

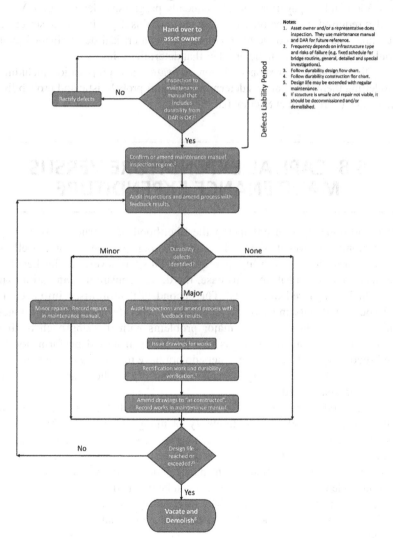

Notes:
1. Asset owner and/or a representative does inspection. They use maintenance manual and DAR for future reference.
2. Frequency depends on infrastructure type and risks of failure (e.g. fixed schedule for bridge routine, general, detailed and special investigations).
3. Follow durability design flow chart.
4. Follow durability construction flor chart.
5. Design life may be extended with regular maintenance.
6. If structure is unsafe and repair not viable, it should be decommissioned and/or demolished.

FIGURE 5.5 Durability process during service life. (From Concrete Institute of Australia, 2014.)

5.7 DETERMINING DAMAGE TOLERANCE

In order to determine damage tolerance and what is a significant level of degradation it is common to use a model with corrosion initiation and then corrosion activation and propagation and consequently progressive deterioration. What is also required from the owner is how much of this degradation the structure can take and still continue to function. This is a critical determination as it could be structural, aesthetic or other that determines this.

Discussion of corrosion initiation, activation and propagation including modelling thereof by various deterministic, semi-probabilistic and probabilistic means is provided at Section 6.8.

5.8 CAPITAL EXPENDITURE VERSUS MAINTENANCE EXPENDITURE

One additional way of enhancing the likelihood of a structure surviving is to monitor it over its design life, and if things are not going as well as hoped, enhance its durability as required. This has several major benefits in that if there is a future court case, the design consultant can claim that they had considered durability. The second is it minimises initial capital spend and puts most of the spend in a separate maintenance budget in the future. There are two major problems with this approach in that firstly, most monitoring systems that are being installed perform poorly and secondly, the future maintainers do not have the money or competence to operate these monitoring systems successfully so the majority of these have been abandoned.

Increased capital expenditure does not necessarily translate into reduced maintenance expenditure. If durability planning is adopted, often capital expenditure is reduced and pretty much without question, maintenance expenditure is reduced. Whole-of-life costs are certainly reduced if a durability planning process is adopted. Maintenance actions and costs are predictable. Accountable maintenance management can be initiated.

Examples have also arisen where 'accounting fudging' of capital expenditure (capex) versus operating expenditure (opex) has occurred by asset owners.

5.9 CONCLUSIONS

The service life of a structure can most certainly be maximised with minimal expenditure (operating and/or capital). There is an appropriately varied range of protection, repair and durability options available so that the maintenance phase in the operational service life of a structure can be optimised at minimal expenditure and whole-of-life costs.

A proactive or reactive maintenance strategy can be adopted with economic impact shown for an asset owner. The proactive strategy normally requires frequent expenditure resulting in an asset that stays in good condition throughout its operational service life, whereas a reactive strategy is accompanied by major rehabilitation/repair costs when poor condition is reached.

Durability planning processes can be developed during the design, construction and operation phases of a structure whether the structure be new or existing. Flowchart examples have been provided.

Determining deterioration tolerance is an important factor and from the asset owner their acceptance of how much of the deterioration the structure can take and still continue to function. This is a critical determination as it could be structural, aesthetic or other that determines this.

It is noted that increased capital expenditure does not necessarily translate into reduced maintenance expenditure. If durability planning is adopted, often capital expenditure is reduced and pretty much without question maintenance expenditure is reduced. Whole-of-life costs are certainly reduced if a durability planning process is adopted. Maintenance actions and costs are predictable. Accountable maintenance management can be initiated.

REFERENCES

Concrete Institute of Australia (2014), Durability Planning, Z7/01 Recommended Practice, Concrete Durability Series, Sydney, Australia, ISBN 978 0 9941738 0 5.

fib (2008), *Bulletin 44: Concrete Structure Management – Guide to Ownership and Good Practice*, February, International Federation for Structural Concrete (fib), Lausanne, Switzerland.

fib (2012), *Bulletin 65: Model Code 2010 – Final Draft*, Volume 1, March, International Federation for Structural Concrete (fib), Lausanne, Switzerland.

Advantages and Disadvantages of Different Remediation Procedures

6

6.1 ANTI-CORROSION PROCEDURE OPTIONS

Extended performance is quite reasonably expected of our reinforced concrete structures. Some, like marine structures, for example, are in aggressive environments where they may be many decades old and of critical importance in terms of function and location or they may be irreplaceable so that engineered repair, protection, maintenance and corrosion management are necessary during their service lives.

Various repair and protection, maintenance and corrosion management approaches are possible during the initiation (t_i), corrosion onset (t_{act}) and propagation periods (t_p) of a reinforced concrete structure. Figure 6.1 presents a summary of the options available, excluding the 'do nothing' option, for carbonation and chloride affected structures.

There are remedial options available that can prevent reinforcement corrosion initiation of a structure such as coatings, penetrants, waterproofing, cathodic protection, electrochemical chloride extraction and electrochemical re-alkalisation.

There are remedial options to slow the rate of reinforcement corrosion (and development of corrosion-induced damage such as cracking, rust-staining,

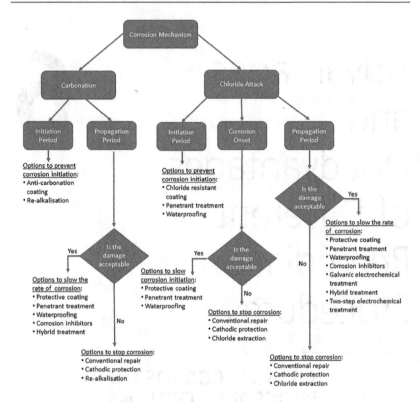

FIGURE 6.1   Flowchart of concrete repair and protection options.

delamination and spalling of concrete). These remedial options include, in no particular order:

- Coatings
- Penetrants
- Waterproofing
- Corrosion inhibitors
- Electrochemical (galvanic anodes)
- Electrochemical (hybrid treatment)
- Electrochemical (two-step treatment)

There are also remedial options to stop corrosion of reinforcement including, in no particular order:

- Conventional (patch) repair
- Cathodic protection (impressed current)

- Electrochemical chloride extraction
- Electrochemical re-alkalisation

Whenever a structure is being investigated all the available options (including 'do nothing') should be presented to the owner for consideration. In short, each option has its pros and cons and the structure owner cannot make a considered decision unless all options are presented.

6.2 'DO NOTHING' OPTION

The 'do nothing' option should always be the first to be considered. It has led to successful long service lives of various reinforced concrete structures. The authors have investigated numerous structures around the world where for large sections the structure owners have not needed to undertake any remedial and protection works to the structures for their over 50 years of service life achieved thus far.

Condition survey work/diagnostic investigations undertaken on the structures then indicate that the structure owners do not need to do anything for the future required service life of the structures.

For marine structures as an example, 'do nothing' also means structure owners not having to complete wharf structure condition assessment manual (WSCAM) inspection and remediation recommendations (Ports Australia, 2014).

6.3 SCENARIO ANALYSES OF OPTIONS

One means of deciding the most appropriate remediation approach(s) for a reinforcement corrosion affected structure is to present a scenario analysis of concrete repair and protection options (including the risks associated with each) to a structure owner. An example scenario analysis for marine wharf substructure elements is provided at Table 6.1. One of the most important aspects which has to be discerned is the client's budget and the likely budget in the future for the residual service life of the structure.

The authors propose that it is also relevant to present to a structure owner a scenario analysis of other remedial options that are available together with the reasons they are not considered appropriate. An example scenario analysis of non-preferred remedial options for marine wharf substructure elements is given at Table 6.2.

6.4 MARINE WHARF SUBSTRUCTURE ELEMENTS – REMEDIAL OPTIONS SCENARIO ANALYSIS

TABLE 6.1   Marine Concrete Substructure Reinforced Concrete Elements – Example Scenario Analysis – Preferred Options

REMEDIAL TYPE	COMMENTS	RISKS
Do nothing, make safe and monitor	• Deflection of deck/beams/pile movements may be the monitoring means. • Diligent and regular inspection required. • Make safe as required. • Local structural failure likely including at 'lime leach cracks'. • Local strengthening (e.g. FRP) may be possible after deflection. • Reactive strategy.	• Localised structural failure. • Reduction in structural capacity and acceptable operation loads. • Re-application of local strengthening (at intervals dependent on strengthening material type). • Highest risk option.
Patch repair and cracks repair	• Proprietary systems. • Conventional techniques. • Concrete breakout of deteriorated concrete. • Breakout and repairs will need to be staged to not affect structural integrity. • Reinforcement to be treated. • Polymer modification and the composition of proprietary cementitious mortars leads to high performance properties (wet and physical) and low penetrability. • Surface coating may be necessary. • Cracks to be widened by saw cutting and breakout. • Crack movement to be determined to ensure structural bond.	• Risk of structural failure since reactive strategy. • Regular structural inspections (every 1–3 years) after initial patch repair. • Patch repairs will need to be ongoing at regular intervals (e.g. every 5–10 years) during life of structure.

(Continued)

TABLE 6.1   (Continued) Marine Concrete Substructure Reinforced Concrete Elements – Example Scenario Analysis – Preferred Options

REMEDIAL TYPE	COMMENTS	RISKS
	• Special design requirements to stop ongoing localised corrosion. • Polymeric crack injection materials not to be used. • Formation of incipient anodes at patch and crack repair locations (but can be eliminated with rebar coatings). • Low capital cost. • Maintenance regime with monitoring required. • Reactive strategy.	• Future service life may become limited and less than desired.
Cathodic protection (impressed current)	• Can completely halt corrosion if correctly designed (regardless of chloride content of concrete, extent and rate of reinforcement corrosion, etc.). • Zoning, sub-zoning, current control, etc. necessary given different micro-exposure environs (tidal, splash, atmospheric), etc. • Likely anode types are mesh, ribbon, discrete or a combination thereof. • Multi-channel transformer rectifier (TR) unit or distributed TR units (power supplies). • Remote monitoring or remote monitoring and control systems can be incorporated. • Reinforcement electrical continuity required. • Less concrete breakout. • Permanent solution. • Routine maintenance (connections, conduits, TR unit modules, etc.).	• Uneven current distribution, uneven protection, acidification issues, etc. if not correctly designed. • Poor component reliability if not correctly designed and specified. • Lowest ongoing risk option.

(Continued)

TABLE 6.1   (Continued) Marine Concrete Substructure Reinforced Concrete Elements – Example Scenario Analysis – Preferred Options

REMEDIAL TYPE	COMMENTS	RISKS
	• Replacement of some components (impressed current systems) every 25 years or so (e.g. TR units, conduits, junction boxes, etc.). • Annual monitoring and reporting. • High initial capital cost.	
Cathodic protection (Galvanic)	• Likely lowest life-cycle cost. • Ongoing cost to maintain. • Trial would be required given resistivity of back of berth lower beam and front beam concrete. • Al-Zn-In metal spray system likely because of higher drive voltage and current output and ease of touch-up. • May not completely halt corrosion but may reduce rate. • Still requires structural monitoring but at less frequent than Do Nothing case. • Reinforcement electrical continuity required. • Permanent solution. • Annual monitoring and reporting. • Routine maintenance commitment. • Anode touch-up repair (and likely replacement) required every 8–12 years. • Anode replacement every 8–12 years (sprayed metal system). • Annual monitoring and reporting. • Lower initial capital cost than impressed current CP but likely higher life-cycle cost than impressed current CP.	• May not halt corrosion but might reduce rate of corrosion. • Structural risk reduced compared to Do Nothing.

6.5 MARINE WHARF SUBSTRUCTURE ELEMENTS – SCENARIO ANALYSIS OF NON-PREFERRED REMEDIAL OPTIONS

TABLE 6.2 Marine Concrete Substructure Reinforced Concrete Elements – Example Scenario Analysis – Other Remedial Options

REMEDIAL TYPE	REASONS NOT CONSIDERED
Surface penetrants/ coatings	• Sufficient chlorides already present to initiate and propagate corrosion at lower section front beam and lower section back beam locations. • Not likely to slow the rate of corrosion (dependent on corrosion kinetics) of lower section front beam and lower section back beam areas. • No need to apply to other substructure beam and soffit areas as chloride build-up within cover concrete is insignificant.
Corrosion inhibitors	• Contentious performance issues (extent of migration; timely migration; locale measurement; longevity; stability). • Accelerate corrosion if insufficient concentration or only partial coverage of the steel reinforcement. • Excessive applications can cause chemical attack of concrete. • In-situ performance monitoring necessary to confirm effectiveness. • Limited performance data most particularly on marine structures. • Mechanisms not understood.
'Hybrid' treatment	• Anode system that is operated in impressed current mode initially (1–3 weeks) and then operates in galvanic mode. • This system is not in the 'Concrete CP Standard' (AS 2832.5-2008). • May not completely halt corrosion but may reduce corrosion rate, therefore may not provide cathodic protection. • Reinforcement electrical continuity required. • Difficult to install (cylindrical type) if congested reinforcement. • Not likely to provide adequate cathodic current given resistivity of lower section front beam and lower section back beam concrete and congested reinforcement. • Recent system, no long-term performance data in a marine environment. • Lack of track record.

(Continued)

TABLE 6.2   (Continued) Marine Concrete Substructure Reinforced Concrete
Elements – Example Scenario Analysis – Other Remedial Options

REMEDIAL TYPE	REASONS NOT CONSIDERED
	• Anodes not easy to find and replace (every 15 years or so) since buried in cover concrete.
	• Anode replacement every 15 years or so (but may be less dependent on anode consumption rates).
	• Trials are recommended given likely corrosion activity of lower section front beam and lower section back beam reinforcement together with the resistivity of the concrete.
	• Trials would need to be undertaken over a number of years so as to also confirm anode output decrease and extent of reinforcement corrosion rate reduction.
Galvanic current devices	• Proprietary, discrete anodes (cylindrical, strip, tubular) embedded within cover concrete.
	• Are not included in AS 2832.5-2008 because do not provide CP per se but rather only some corrosion rate reduction.
	• Not likely to provide adequate cathodic current given resistivity of lower section front beam and lower section back beam concrete and congested reinforcement.
	• Difficult to install if congested reinforcement.
	• Anodes cannot be replaced in the future.
	• Anodes will attract chloride ions into concrete and may be source of localised corrosion risk to reinforcement when anodes consumed.
	• Trials would need to be undertaken over a number of years so as to confirm anode output decrease and extent of reinforcement corrosion rate reduction.
	• Reinforcement electrical continuity required.
Electro-chemical chloride extraction (desalina-tion)	• Application constraints.
	• Ongoing maintenance costs (i.e. chloride barrier coating).
	• Not all chlorides can be removed.
	• May cause ASR of concrete or change its morphology.
	• Routine maintenance (e.g. touch-up, etc. of chloride barrier coating).
	• Local concrete patch repairs.
	• Re-application every 10–15 years of chloride barrier coating.
Replace-ment	• Impact on operations.
	• Berth (or sections of berth) would need to be closed.
	• High cost.

6.6 DECIDING ON THE MOST APPROPRIATE REMEDIATION PROCEDURE

Decisions on remediation procedures for reinforcement corrosion affected concrete structures should be based on knowledge of:

- The cause(s) of deterioration
- The degree and extent of deterioration
- The expected progress of deterioration with time
- The effect of the deterioration on structural behaviour and serviceability
- The money available
- The level of disruption to operations allowable

Knowledge of where corrosion is occurring, together with information regarding the causes of corrosion, is always necessary before deciding on the most appropriate, or combinations of appropriate, remediation procedures.

Other factors that need consideration include, but are not limited to:

- Structure type
- Element types
- Future service life

After the conduct of the above, deciding on the most appropriate remediation procedure should then be undertaken using, for example, the flowchart approach and scenario analysis approaches presented earlier.

As an asset owner or other stakeholder, it is strongly recommended that conduct of the above, including presentation of remediation options, be undertaken by independent consultants and not consultants that are conflicted nor by contractors (who are obviously conflicted) or organisations that wear multiple hats (e.g. consultants/suppliers/installers).

6.7 CONDITION SURVEY TO ASSESS THE CAUSE, DEGREE AND EXTENT OF DETERIORATION

A condition survey and representative testing of a structure/elements can be undertaken to assess the cause, degree and extent of deterioration. For reinforcement corrosion-induced deterioration, a condition survey and representative testing will enable determination of where corrosion is occurring together with information regarding the causes of corrosion.

A condition survey is a process whereby information is acquired relating to the current condition of the structure with regard to its appearance, functionality and/or ability to meet specified performance requirements with the aim of recognising important limitations, defects and deterioration (Concrete Institute of Australia, 2014). A wide range of parameters may be included within a condition survey with data being obtained by activities such as visual inspection and various forms of testing. A condition survey would also seek to gain an understanding of the (previous) circumstances which led to the development of that state, together with the associated mechanisms causing damage or deterioration (Concrete Institute of Australia, 2014).

Discussion of relevant aspects of various condition surveys of concrete structures is beyond the scope of this short book other than the discussion of NDT techniques for detecting and monitoring corrosion provided in Chapter 3 earlier.

It should be noted that in order to obtain representative test results, the locations of on-site testing, sampling and laboratory testing must be representative of the population. In determining the locations to be tested, due consideration should be given to the configuration of the structure/element, the macro and micro exposure and the method of construction (Concrete Institute of Australia, 2015). When developing the testing plan, the Concrete Institute of Australia (2015) proposes that the item to be assessed (structure or exposure) should be split into zones where the results can reasonably be expected to be similar. In that way they advise that the sample results expressed as a mean and variance will give a true statistical representation of the zone.

Emphasis should be placed that a condition survey (as well as consequent presentation of remedial options) should again only be undertaken by independent consultants and not consultants that are conflicted nor by contractors or organisations that wear multiple hats.

A condition survey may also be referred to as a condition assessment and/or a condition evaluation. It is considered worthwhile that the Concrete Institute of Australia (2014) terminology for condition assessment and condition evaluation be reproduced here:

- **Condition Assessment**: a process of reviewing information gathered about the current condition of a structure or its components, its service environment and general circumstances, whereby its adequacy for future service may be established against specified performance requirements for a defined set of loadings and/or environmental circumstances.

- **Condition Evaluation:** similar to a condition assessment, but may be applied more specifically for comparing the present condition rating with a particular criterion, such as a specified loading. Condition evaluation generally considers the requirement for any later intervention which may be needed to meet the performance requirements specified.

An assessment of the remedial options that are relevant to different parts of a structure or its elements is also often a part of a condition survey report. As indicated previously, the flowchart approach and scenario analysis approaches presented earlier are then considered relevant.

6.8 REINFORCEMENT CORROSION MODELLING

During the lifetime of a reinforced concrete structure, it has been proposed that there are two basic periods with respect to corrosion and corrosion protection of reinforcement. The initiation period (t_0), when critical amounts of chloride, when carbonation or when leaching has not reached the reinforcement so that the reinforcement remains passive, and the propagation period (t_1), when reinforcement corrosion proceeds (see Figure 6.2(a)). This corrosion model ($t_0 + t_1$) was first proposed by Clear and has since been developed by many others since 1974.

The conventional model of Clear (1974) or Tuutti (1982) (Figure 6.2(a)) for the commencement and development of reinforcement corrosion has negligible

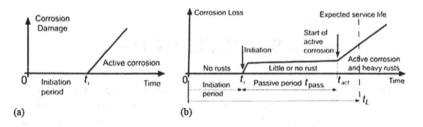

FIGURE 6.2 (a) Classical Tuutti (1982) model showing no corrosion (damage) until t_i followed immediately by development of serious corrosion (damage); (b) modified model with separation between initiation at t_i and start of active corrosion at t_{act}. (From Melchers and Li, 2006.)

or no corrosion before the initiation time (t_i), immediately after which corrosion becomes serious. In terms of the development of chloride-induced reinforcement corrosion models since Clear and Tuutti, Melchers and Li (2006), for example, have proposed that some limited amount of corrosion may commence at the initiation time t_i but that serious long-term corrosion generally does not commence until some later time t_{act} (see Figure 6.2(b)). In this model, the period (0-t_i) has essentially zero or negligible corrosion. This is followed, commencing at the initiation time t_i, by some corrosion for a relatively short period of time, after which there is no, or only a very modest, increase in corrosion, until the commencement, at the activation time t_{act}, of serious, damaging corrosion. Melchers and Li (2006) proposed that the corrosion process in the time period 0-t_i differs from that causing the serious long-term corrosion.

The time to corrosion initiation (t_i) for carbonation and chloride-induced reinforcement corrosion can be modelled by various deterministic, semi-probabilistic and probabilistic means and is reasonably well established. Reinforcement corrosion propagation modelling, on the other hand, is not well established.

To determine the time for the corrosion propagation period, two major parameters need to be determined. One is the critical criterion for deterioration damage by reinforcement corrosion. The other is the corrosion rate of reinforcement in various conditions. Once the two parameters are determined, the length of the reinforcement corrosion propagation period can be determined.

The criterion for deterioration failure of a reinforced concrete structure depends on the permissible deterioration level for the limit states specified by the asset owner or any other stakeholders.

The corrosion rate of reinforcement in concrete very much depends on the environment conditions, cementitious materials composition, the concrete quality and the thickness of cover concrete. The corrosion rate also depends on the type of depassivation, i.e. carbonation or chloride penetration.

6.9 CONCLUSIONS

Extended performance is often expected of our reinforced concrete structures. Some, like marine structures, are in aggressive environments, they may be many decades old, of critical importance in terms of function and/or location or be irreplaceable, such that repair and protection is necessary during their service lives.

Condition surveys are an essential part of the management and maintenance of reinforced concrete assets.

Various repair and protection, maintenance and corrosion management approaches are then possible during the initiation (t_i), corrosion onset (t_{act}) and propagation periods (t_p) of a reinforced concrete structure.

The 'do nothing' option should always be the first considered. It has led to successful long service lives of various reinforced concrete structures where for large sections the structure owners have not needed to undertake any remedial and protection works for their over 50 years of service life achieved thus far.

Remedial options are available that can prevent reinforcement corrosion initiation, that can slow the rate of reinforcement corrosion and, in some limited cases, stop corrosion of reinforcement. The number and type of options mean that where structures and buildings may be at, or beyond, their design lives, many decades of future service lives can be pro-actively engineered at minimum lifecycle costs.

REFERENCES

Clear K C (1974), Evaluation of Portland Cement Concrete for Permanent Bridge Deck Repair, Report FHWA-RD-74-5, Federal Highway Administration, Washington, DC.

Concrete Institute of Australia (2014), Recommended Practice, Concrete Durability Series, Z7/01 Durability Planning, Sydney, Australia.

Concrete Institute of Australia (2015), Recommended Practice, Concrete Durability Series, Z7/07 Performance Tests to Assess Concrete Durability, Sydney, Australia.

Melchers R E and Li C Q (2006), Phenomenological modelling of corrosion loss of steel reinforcement in marine environments, *ACI Materials Journal*, 103, 1, 25–32.

Ports Australia (2014), Wharf Structures Condition Assessment Manual, Sydney, Australia.

Standards Australia (2008), AS 2832.5–2008 Cathodic Protection of Metals - Part 5: Steel in Concrete Structures, Sydney, Australia.

Tuutti K (1982), Corrosion of steel in concrete. In S. C. Institute, CBI Research Report no. 4.82. Stockholm, Sweden: Swedish Cement and Concrete Research Institute.

Examples of Damage and Remediation with Different Structures

7

7.1 MARINE JETTY

7.1.1 Jetty Description

The original inner walkway dates from around 1979 with reinforced concrete vertical and raking piles from deck level to below the water level (Figure 7.1). The entire structure was cast in-situ reinforced concrete. Extensive corrosion damage to the steel reinforcement was observed at many locations. This was particularly evident on the soffit and above high-water level on the rectangular section lattice substructure. A significant repair program has been undertaken at a previous time at above the low tide level. In this repair, cracked concrete has been removed and replaced with newly cast concrete. This program did not stop the corrosion and some of the repairs have now also failed due to ongoing oxidation of the steel reinforcement. No evidence of structural distress was evident.

The outer walkway was built in 1995 using steel piles combined with the upper structure of reinforced concrete (Figure 7.2). The steel piles are protected up to water level by galvanic anodes and above by a coating. The pile heads are precast concrete and the walkway is a combination of precast planks and in-situ kerbing. All of these various reinforced concrete structures are showing some degradation due to corrosion of the rebar. This was particularly evident where there is limited cover depth.

FIGURE 7.1   Columns, beams and rakers on inner wharf.

FIGURE 7.2   Outer precast section.

7.1.2 Life Extension Options

7.1.2.1 Do Nothing

The structure is presently performing its function. That is carrying pipelines, providing a walkway and berthing for ships and is likely to do so for a considerable future period. If a residual life can be decided and it is relatively short, say the next five years, then the risks of the continuing corrosion degradation causing structural distress are fairly low and it is reasonable to do nothing.

7.1.2.2 Do Nothing and Regular Inspections for Structural Stability

If a residual life requirement of greater than five years is required then a regular inspection regime can be instituted. This regime does not have to follow the commonly prescribed testing procedures that are in the recommended practices for assessing reinforced concrete structures, such as half-cell (electrode) potential mapping, chloride samples and drawings showing deterioration. This is because you already know that the problem is that the steel reinforcement is corroding. You also know where the worst damage is occurring because you can see cracks, missing concrete, staining. The only thing you need to be actively interested in is where and when this deterioration is leading to structural distress causing the functionality of the jetty to be reduced. To give an example, the reinforced concrete of the walkway deck could be cracking because the lower jetty is not carrying the load adequately. This then is increasing the point loading and causing damage to the walkway. This walkway could then become unusable with consequences on the functioning of the structure.

The structural stability survey can be quick and simple with a boat and walkway trip as all that is being looked at is structural distress.

7.1.2.3 Patch Repairs

This was undertaken in the past on the inner jetty. If undertaken now, it is going to be fairly expensive (to repair most of the cracked or delaminated areas on the structures will require a million-pound investment).

7.1.2.4 Structural Augmentation

This could be undertaken in conjunction with regular inspections for structural integrity. When signs of structural distress appear then you strengthen these areas. For example, this could take the form of additional piling if it is the columns tied into the structure or RSJ's bolted to the failed rakers or beams. It could also take the form of post-tensioning tendons being applied along the deck. Contrary to normal bridge-strengthening applications this installation does not have to have aesthetic virtues and this makes it relatively simple and cheap to implement.

7.1.2.5 Cathodic Protection

If you want to stop the corrosion of the steel reinforcement then you could put an impressed current cathodic protection (ICCP) system on some or all of the structure. This type of installation is customised for atmospherically exposed concrete and has little in common with the existing galvanic system which is protecting the in-water steel piles of the structure. The problem is cost. The cathodic protection anode system which is most logical for this structure is drilled in anodes on the columns, rakers and pile caps. MMO ribbon anodes will be most suitable on the slabs. Typically, in the UK there would be an installed cost of around £200 a square metre for a complete ICCP system with anode, monitoring system and power supplies. Due to the poor access, tides and likely poor continuity (the precast items will all have to be made electrically continuous), the price could easily be £400 a square metre. Over a total area of 6,000 m² this would give a ballpark figure of £2.4 million. It will also have to be maintained. An alternative to this is to only ICCP areas which are perceived to be most at risk of structural failure. This procedure would require the services of a structural engineer who is experienced in the behaviour of this type of ageing structure to select the most appropriate areas.

7.1.2.6 Other Electrochemical Treatments

Another thing you can do is remove the chloride ions from the concrete by electrochemically moving the chlorides into a buffered solution on the concrete surface. The surface is then coated as the concrete porosity will be increased by this process. The cost of this is likely to be substantially more than ICCP. There is also a significant level of debate about the effectiveness of this process in preventing further ongoing corrosion on structures such as this. One concern is whether the chloride which has penetrated further than the first layer of reinforcement is included in this removal circuit. You cannot physically remove the chloride by knocking out the contaminated concrete as none of the beams, columns and rakers would be left. Galvanic anodes, hybrid treatment and two-step electrochemical treatment would have limited effectiveness on a structure such as this with high corrosion rates.

7.1.2.7 Replacement of the Structure

This is building a new structure alongside the existing jetty and wharf, then transferring all the services onto the new structure.

7.1.3 Discussion

The two most relevant aspects are the intended residual life expectancy of this structure and the funds available now and in the future.

The prevalent form of deterioration observed was chloride-induced corrosion of the steel reinforcement. This is likely to be the life determining deterioration mechanism for this structure.

If the structure is likely to be required for an extremely long time then a significant financial investment will be required for one of the three options:

- To reduce (stop) the rate of corrosion – the most cost-effective way of doing this is ICCP.
- Structurally add or replace parts of the structure.
- Build a completely new jetty and wharf alongside the existing structure.

The lowest capital cost option for the next 20 years is to replace the parts of the structure or strengthen it as is structurally required, and this was the route chosen by the client along with structural distress surveys on a biannual basis.

7.2 PRESTRESSED CONTAINMENT TANKS

Four tanks were constructed in 1980 in a coastal environment to act as emergency containment bunds to main storage tanks. These tanks are comprised of cast in-situ reinforced concrete panels around which a stressed wire is spun. The wire is covered by a sprayed concrete (known at this time as gunite) which provided part of the corrosion protection of the wires which were also galvanised. There are 14 prestressed bands of reinforcing wire installed around the tank. The post-tensioning wire was nominally 5 mm in diameter and was wrapped under tension by a machine hung from wheels running around the top of the structure in bundles of around 100 wires to make up these bands. The tension applied to the wire was around 60% of its ultimate tensile strength.

In 2012 a failure of one 'ring' of a metre or so section of the pre-stressing wire and gunite overlay occurred around the complete periphery of the tank (Figure 7.3).

An examination of the pre-stressing wire in this band showed that corrosion had occurred in a significant proportion of it. There are two steel elements of the structure which could corrode, the first being the steel reinforcement in the cast in situ wall. This is different to the post-tensioning wire in that it does not have a protective zinc coating but is completely encased

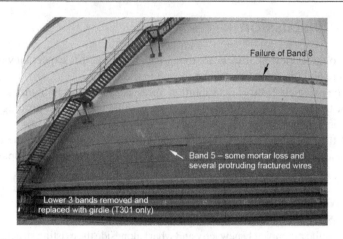

Failure of Band 8

Band 5 – some mortar loss and
several protruding fractured wires

Lower 3 bands removed and
replaced with girdle (T301 only)

FIGURE 7.3 View of a tank after band failure.

in concrete. This type of construction can suffer from significant levels of corrosion without loss of structural integrity. This inner concrete structure is protected from the atmospherically borne chloride by the banding and showed no signs of distress so all remedial efforts were made considering the wire and gunite layer.

The tanks closer to the sea and also the sides facing the prevailing winds (westerly) showed more signs of corrosion which made it apparent that airborne chloride was the catalyst for this corrosion (Figure 7.4). Particular damage was noted where water had pooled on ledges.

7.2.1 Bund Repair Options

There are several approaches that can be taken with these structures apart from complete demolition and replacement. All of them assume that there is no corrosion to the cast in-situ part of the bund. This was verified by a visual survey, followed by selective pits (breakouts) dug through the cover to allow visual inspection of the reinforcement. A half-cell (electrode) potential survey

FIGURE 7.4 Severe corrosion of the post tensioning wire.

at high risk areas was not undertaken as there would be confusion over the presence or absence of the zinc coating.

Strengthening of the outer layer will be required, as there is no way of knowing how much strength has been lost by the post-tensioning wire. Since this is the case the question of whether the bund is fit for purpose cannot be answered until it is used in earnest. This is obviously unacceptable.

Some of the alternative options are given below:

1) Removal and replacement of the existing post-tensioning wire and gunite with new reinforcement and gunite is straightforward and direct and will give an assured life expectancy.
2) Place post-tensioning wire directly over the existing post-tensioning wire and gunite. For this option to be valid an assessment on whether the crush strength of the inner wall is exceeded would need to be undertaken. This is because the additional post-tensioning will increase the stress on the cast in-situ inner wall.
3) On the so far undamaged part of the structure, modifications to reduce the corrosivity of the post-tensioned wire environment by the use of cathodic protection. The premise that there is no ongoing corrosion could be validated by acoustic emission as well as traditional electrical sensors.

7.2.2 Discussion

There was a significant amount of corrosion on the post-tensioned wire which had resulted in partial structural failure on the outer part of the bund. This failure occurs with this type of construction detail as compared to the more common unstressed reinforced concrete which can tolerate significant to extreme corrosion damage before structural failure. The time that has elapsed since construction means that there will be more and more corrosion-induced damage occurring on all of these structures in the immediate future.

Sophisticated non-destructive inspection techniques to determine the extent of the corrosion were considered. It was considered to be more cost-effective, simpler and more reliable to combine a visual inspection with cutting out pockets of gunite for direct visual inspection of the wires.

The client eventually decided on total demolition of all the bunds and replacement with a similar design with the addition of a high build coating.

7.3 MIDDLE EASTERN AIRPORT

7.3.1 Introduction

A very large new extension to an airport was being planned which essentially comprised a giant reinforced concrete box to be placed under the concourse which was to house a massive baggage-handling system and other facilities. Groundwater would cause this huge construction to have buoyancy so the piling system was designed for tension. The life expectancy that was required in the specification was more than 100 years. In the Middle East in the 1970s little thought about long-term durability had been given and standard Northern Europe and United States designs were used with no modifications for the particularly aggressive conditions that appertained. After widespread problems at all the coastal locations around the Gulf with premature deterioration of bridges, tunnels, marine facilities and cooling towers, to name but some of the facilities, this attitude changed. At the time of this extension being designed in 2006 the adjacent earlier airport below-ground structures were showing significant signs of premature reinforcement corrosion. This required several extensive, retro-fitted, ICCP installations to reduce the ongoing rate of degradation. This premature degradation caused the designer to consider more carefully what course of action would be required to achieve 100 years of life and not just use the same design as in previous airports.

7.3.2 The Problem

The Middle East is hot and in the Gulf region there is a marine atmosphere and the groundwater is also chloride contaminated. The soil is normally slightly compressed coral sand which has a high chloride content. It has been found in all coastal Gulf States that very significant corrosion problems, particularly of tunnels and other below ground level structures, were occurring because the inner surfaces of these are dry and hot while the outer surface has chloride containing water which can be pressurised by the depth below water level.

Due to the very much smaller pore size of the concrete compared to the sand matrix, water is drawn (Hall and Hoff, 2012) from the soil and fully saturates the structure. This concrete is significantly wetter than the surrounding soil. There is then capillary movement of the water and salts through the structure. The chloride ions are then moved through the concrete by advection. On the inner warm dry surfaces of the structure, water evaporates creating a water flux through the concrete. As the water evaporates the chloride is left behind and increases in concentration. This, coupled with the higher oxygen concentration on the inner surface, starts corrosion of the reinforcement at the lowest cover depths.

On the older airport structures any weaknesses in the integrity of the build, such as cold joints or poor compaction, showed the first signs of corrosion such as staining. This is because this advection and capillary process carried the soluble corrosion product ions to the surface where further oxidation gave an insoluble salt and brown staining.

7.3.3 The Solution

The designer considered several options such as stainless steel reinforcement, ICCP from large groundbeds beneath the slabs and ICCP as ribbon anode on the inside layer of the reinforcement but was deterred by the cost and delays to the construction schedule these options would require. Instead they decided on a liquid membrane form of tanking on the bottom surfaces of the slabs and walls combined with a corrosion monitoring system.

7.3.4 Previous Corrosion Monitoring Systems

The purpose of a corrosion monitoring system was to determine if parts of the structure are actively corroding. Previous corrosion monitoring systems on large reinforced concrete structures have been installed with varying levels of success (Chess, 1993) in achieving this result. Typically, these monitoring systems are looking at the electrochemical state of the steel or changes in the resistance of the concrete. They have two major problems.

Firstly, they require real time calibration. This is because, for example, the data you are getting is that the potential of the steel has changed over a period of time. The reason for this change could be a change in the reference electrode composition, change in circuit dynamics or change in oxygen level at the surface of the steel reinforcement. So you cannot say with any level of certainty that as the potential becomes more negative, corrosion is now occurring. Thus, you also need a verification procedure such as cracking or staining or a destructive inspection which may not be practical and requires a significant maintenance cost.

The second problem with these monitoring system types is that they are only relevant in the immediate location where the sensors are located. For example, a reference electrode will be addressing the electrically nearest 20–50 cm of steel reinforcement surrounding the sensor. This is the same problem with all the sensors which can be directly buried in the concrete and it means that only a tiny percentage of the structure is actually being monitored. This may still be a reasonable approach if the chloride is travelling on a broad front from the outside of the structure to the inside, but this is not what was actually found in earlier concourses where cold joints allowed rapid localised ingress. Because of these two factors a new monitoring system concept was deemed necessary.

7.3.5 New Corrosion Monitoring System

The thick bottom slab (typically 3 m) was cast onto blinding which had the waterproofing membrane applied to it and the pile cages feeding in from below. Above this a 'sanitary fill' was added, which is sand, and then a relatively thin floor slab was cast on top. See Figure 7.5 for a schematic of this arrangement.

The reason for this was originally that the level of the floor slab was uniform but it had some important consequences. Firstly, there were watertight areas each of about 500 m² which would fill with groundwater if cracking of the bottom slab or advection occurred. In total there were 270 of these floor slab areas in the structure.

For durability a hydrostatic sensor was chosen with a duplex stainless steel body, an alumina ceramic diaphragm and a viton seal. These were placed in a perforated plastic tube. A pair of these sensors were placed in each of the separate slab areas, with appropriate electronics and the copper and fibre optic wiring all run to a control computer in the central room as shown in the below schematic in Figure 7.6. In this control room the computer would be updated

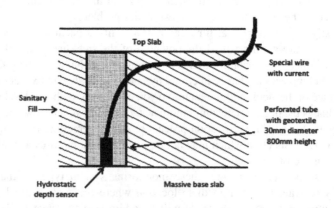

FIGURE 7.5 Arrangement of water sensor on top of base slab.

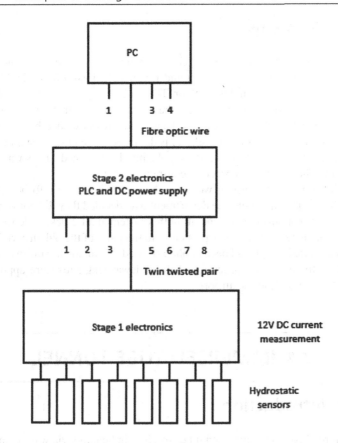

FIGURE 7.6 Schematic of wiring arrangement.

every 10 minutes with the depth of any liquid in all the floor slab area and alarms set with a full record kept.

This unusual arrangement had several immediate and important advantages over previous monitoring systems applied to other structures.

Firstly, it was monitoring almost all the area of the base slab (approximately 135,000 m^2) and secondly, there was a very simple response to a sensor warning. If the sensors indicated a liquid level above the base slab then you went to the location identified, drilled a hole through the top slab and took a water sample. This sample could then be analysed to define whether the liquid was groundwater or a leak from an airport pipeline. With this information the appropriate remediation could be devised, such as installing a local ICCP system for the flooded base slab or repairing the broken water pipe.

7.3.6 Problems

The first large problem was that this corrosion monitoring system was included as part of the groundworks contract and was installed early in the build with the cabling run in ducts under the floor. There was many kilometres of cabling for this system in both the floors and the walls. Later contracts for fitting all the services, particularly the baggage-handling system included bolting many large devices to the floors and walls. This damaged some of the system wiring which caused significant problems as replacing the damaged cables was complicated by the verification process required.

With most of the system working, the system was ready to be handed over to the airport maintenance department who decided they did not want the expense and responsibility of operating this system at all. Luckily, dewatering for a further extension to the airport close to the parts of the old airport facilities where retrofit ICCP had been applied showed a dramatic fall off in damage caused by the advection process and these dewatering sites were applied at more locations around the airport.

7.4 IMMERSED TUBE TUNNEL

7.4.1 Introduction

A reinforced concrete immersed tube structure in Scandinavia was completed in 1999. The segments of the tunnel were made in a precasting yard, floated out and sunk into a seabed trench. The segments were connected together on the seabed with a sealing system which was supported using a steel frame and the structure covered with sand and overburden so that the original seabed profile was maintained. Unlike older designs of this tunnel type, no anticorrosion measures such as cladding, coating, ICCP to the outside of the segments were deemed necessary, beyond high-strength concrete and a large cover depth, to achieve its 120-year design life.

After the twenty segments were built but before they were floated, it was discovered that seven of the end frames were inadvertently electrically connected to the tunnel reinforcement during the construction process. These end frames were considered to be at risk from accelerated corrosion due to galvanic interaction with the reinforcement in the concrete segments. This is because there is a huge surface area of steel in concrete which has a less negative potential than the steel end frame in seawater. It was advised that this corrosion could constitute a threat to the structure's integrity in the middle term and the client made the contractor take additional anti-corrosion measures.

To prevent this occurrence of galvanic interaction a galvanic cathodic protection system for each of the end frames was designed and installed after the tunnel was laid in the sea trench but before backfilling. With this galvanic zinc anode system only two of the seven end frames achieved a satisfactory protection level as measured by reference electrodes on the outside of the tunnel. Thus it was later decided to apply an ICCP system to the five end frames. This system was installed in autumn 2003. Since this time a significant amount of monitoring (including diving and removal of the overburden) has been undertaken on these cathodic protection installations and what they are achieving on the structure. The results have been ambiguous at a very high cost to the owner running into the millions of euros.

7.4.2 Appropriate Solution?

A drawing of a typical sealing frame is given below in Figure 7.7. This diagram is probably not exactly correct, as on this project it has been stated (Lunniss, 2013) that the counter plate assembly was dispensed with, presumably to save money. So there is only one end frame on each segment and not a male and female.

Figure 7.8 shows a segment being lowered to the seabed with part of the sealing system (Gina seal) between segments evident.

What is apparent from Figure 7.7 is that only a small part of the seal assembly can actually receive cathodic protection from external anodes

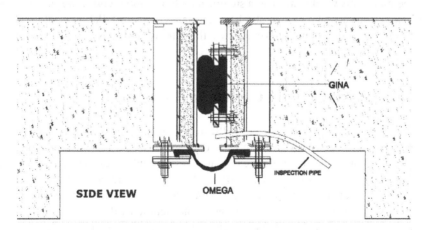

FIGURE 7.7   Roof of immersed tube showing sealing arrangement.

FIGURE 7.8 Segment being lowered to the seabed showing the continuous Gina seal.

irrespective of whether it is a galvanic or an impressed current system. This is because all the metal work below the Gina seal is electrically isolated from the seawater through which the current flows and the backing plate will prevent cathodic protection current indirectly coming through the concrete onto the steel between the Gina joint and the Omega seal. This electrical isolation also means that there is only a small part of the entire joint detail that is actually at risk of enhanced corrosion.

A much cheaper and more effective solution would have been to provide a di-electric shield for all the metallic parts of the joint which were at risk of this enhanced corrosion rather than the cathodic protection adopted. This could have taken the form of bonding rubber sheeting to the steel around the outer bolts of the Gina seal and then outwards over all the coated metal structure of the joint. This would have had a significantly lower initial cost, lower maintenance requirement and a greater life expectancy. This installation would still be effective after many years even when the adhesive failed as it would be held in place by the backfill.

7.4.3 What is the Cathodic Protection Doing?

Recently a paper (Klinghoffer, 2018) discussed the performance of these cathodic protection systems and the data given will be used. According to Klinghoffer the theoretical calculation of current necessary for protection of a single end frame was estimated to be approximately of the order of 0.6A. In common with many galvanic anode installations there was no provision for current monitoring and so no detailed information on their respective outputs is available, but the actual output was significantly more than the estimate and was in the region of 5A. For the impressed current

TABLE 7.1 Output Current and Protection Levels for the ICCP System

END FRAME ID	$V_{INSTOFF}$ (MV VS AG/AGCL)	CURRENT (A)	VOLTAGE (V)	ASSESSMENT OF PROTECTION
EF2	−1032	24	5.1	overprotected
EF4	−1023	7	3.1	overprotected
EF5	−450	43	7	lack of protection
EF7	−650	35	6.6	lack of protection

system more data was available for the anode array and this is given in Table 7.1.

This data shows huge variations in the output level and performance of the cathodic protection systems. This would not normally be expected to be occurring, as the design of the impressed current systems was the same and the environment surrounding the tunnel segments is also the same. The protection current of the CP systems is clearly going to somewhere different to its design target of the end frames. In some cases, such as EF5 and EF7, even at maximum power supply output of the CP system, there is insufficient protection to the end frame in order to stop corrosion. It should be noted though that this is a very different requirement to that of preventing the galvanic effect which was the original purpose of the cathodic protection.

To further investigate the situation, reference electrodes which were installed in the tunnel wall at a location 170 m from an impressed current system and also about 80 m between two locations with ICCP showed significant changes with the interruption of the current to the impressed current anodes. There was significant cathodic protection being measured on the steel reinforcement in the top outer walls of the tunnel. There was a much smaller effect measured by a reference electrode at the bottom outer wall of the tunnel.

This cathodic protection may be helpful to improving the durability of the tunnel as it will supress any corrosion of the steel reinforcement, particularly on the outside of the tunnel walls. In this combined rail and road tunnel the biggest future corrosion problems will probably be de-icing salt carried into the tunnel by vehicles on the road tunnels and AC and DC stray current corrosion in the train tunnels. For these potential problems, the external CP systems will at best only have a marginal effectiveness at certain limited locations.

When the galvanic and impressed current systems were designed it was not understood that the individual segments have two other joints with a possible electrical discontinuity at both these locations. These joints are match-cast segment joints that have no continuity reinforcement through them and are designed to allow differential settlement to be accommodated. They are grouted up during the final stages of tunnel completion.

To make matters even more complex, there are many electrical installations throughout the tunnel and they are commonly earthed to the steel reinforcement. Then there is the traction current for the trains. The way the electrical systems have been wired and how common or isolated individual sections of the tunnel walls are is the most probable reason for the very different protection levels for the similar CP systems.

7.4.4 Discussion

The CP systems appear to have been installed as a punishment for the contractors' failure to maintain electrical isolation of the end frames during construction in the dry dock. The concern about galvanic interaction could have been addressed by electrical isolation of the steelwork using rubber or a similar material at a much lower installation cost and no maintenance requirement. Once a cathodic protection system was installed there was 'mission creep'. Initially its purpose was to prevent this enhanced galvanic corrosion. Then the purpose changed to stop any corrosion of the end frame and now it is protecting the reinforcing steel on the outside of the tunnel as well as the outer part of the end frame from corrosion.

Now the tunnel has been in operation for 20 years it is logical for the owner to look at all the likely durability issues that might be reasonably expected to endanger the designed life expectancy. For example, on another older immersed tube tunnel in Britain they have had substantial roof cracks with leakage of salt water and ferric oxide staining indicating active corrosion, vulnerability of the Omega seals to fire damage and concerns about de-icing splashing causing corrosion of the steel reinforcement in the tunnel side walls. In this tunnel they installed a corrosion monitoring system during construction which proved too complex to provide any useful data on the actual structural durability.

7.5 SWIMMING POOL

7.5.1 Introduction

Chiltern swimming pool was constructed in 1972 in England on behalf of the local council. The pool has since become a very popular facility in this town. The pool building comprises a steel portal framed building with in-situ concrete pool basins and poolside walkway slabs. Access is made available to the dry side of the pool basins via a poolside service void.

The main pool basin and the pool building frame have experienced significant structural decay over their life until 2007, i.e. 25 years, most notably in the form of spalled concrete due to expansive corrosive reactions occurring in the reinforcement and corrosion at the base of the steel stanchions (Figure 7.9).

The majority of the areas of damage to the pool basin are located in areas of inadequate concrete cover to the steel, which was noted to be zero over significant areas (Figure 7.9). It is believed that the lack of cover was the result of poor quality control during the construction of the pool. This lack of quality control was confirmed by the large amounts of formwork timber left encased in the concrete and the large number of steel formwork bolts and brackets left protruding from the concrete surfaces.

The precast concrete building frame would be expected to have been built with a higher level of quality control than the in-situ components. Unfortunately, however, the building frame was not designed specifically for use in swimming pools with only 25 mm of concrete cover having been provided to the steel (mild exposure conditions for a grade C30 concrete). In reality the exposure conditions within the poolside service void are very severe.

In addition to the poor concrete construction quality, significant areas of the tile grout and the expansion and contraction joint sealants around the pool have failed. This has resulted in a steady flow of chlorinated water over the exposed concrete faces (Figure 7.10).

FIGURE 7.9 Areas of spalled concrete on the main pool basin wall. The very low cover to the steel reinforcement can be seen.

FIGURE 7.10   Encrustation of the main pool basin wall. These deposits are signs of advection.

A full structural appraisal of the pool building frame and the pool basins in the damaged state to assess the risks to the public and staff at the leisure facility was undertaken. The levels of reinforcement were determined following inspections of the areas of spalled concrete. It was found that the reinforcement provided in the pool basin was the minimum area of reinforcement and was not required for the long-term stability of the pool basin. The pool basin was shown to be of adequate strength and stiffness even allowing for a large loss in cross-section due to the spalled concrete. The building frame was assessed with the assumption that a plastic hinge had formed at the location of the severe spalling. The frame was also shown to be stable in its damaged state. Despite having shown that the structure was stable in the ultimate limit state it was concluded that concrete repairs should be undertaken to help prevent further decay and serviceability problems in the future.

Half-cell (electrode) potential and cover meter surveys were undertaken and concrete samples were taken from a number of locations around the pool basin and the frame. The samples were tested for chloride content, carbonation, sulphate and concrete strength. It was shown that sulphate levels were well within the prescribed limits, the carbonation levels were low and that the concrete was of a reasonable quality. The chloride content results, however, showed a large variation across the samples tested with a number demonstrating very high levels of chlorides, which was consistent with the half-cell (electrode) potential survey.

It was concluded that the main factor affecting the decay of the structure was the high level of chlorides due to the penetration of swimming pool water into the concrete through the advection process (Figure 7.10).

Given the importance of the facility it was considered vital that any proposals for remedial works provide a robust long-term solution. Consideration was given to a number of options:

- Patch repairs to concrete – The concrete would be repaired over the damaged areas only. It is possible that in future years additional repairs would be required as the building continued to decay.
- Chloride extraction – Consideration was given to the extraction of the chlorides from the concrete. It was considered that due to the high risk of future penetration of chlorides into the concrete, this would only provide a temporary solution and would need to be repeated in future years.
- Cathodic protection – Given the significant level of chloride penetration and likely further advection, it was concluded that the only satisfactory long-term solution for the repair of the pool basin and pool building was the patch repair of damaged concrete followed by the application of an impressed current cathodic protection system to the areas at risk. The ICCP system was specified to have a design life of 25 years minimum.

7.5.2 Selection of a Cathodic Protection System

The typical anodes for ICCP on most reinforced concrete structures are applied to the structure as a coating, mesh or drilled-in anodes, as described in Chapter 6. This is expensive and difficult to achieve due to the confined nature of the tunnel and large amount of metal supports in the wall so an alternative was sought.

It was found that a Danish company called Cathodic Protection International ApS (CPI) had, since 1982, been supplying cathodic protection systems which used anodes in the pool water. Since then CPI has supplied cathodic protection for more than 60 swimming pools in Denmark, with at least 50 of them computer-controlled. Other countries have discovered the benefits of cathodic protection of swimming pools, and CPI has supplied these systems to Norway, Sweden, Luxembourg, France and the UK among others.

The cathodic protection of a swimming pool can consist of two different methods:

- The first method is only suitable for the pool structure and involves anodes placed in the water, which are called water anodes. These are normally recessed into the wall and fixed in plastic boxes sized to replace one or two tiles. The current is passed through a perforated lid to the water, from the water through the grout between the

tiles and through the concrete to the reinforcement steel. A 25 m ×
12 m swimming pool would normally require 6–12 water anodes.
Water anodes are also applicable for concrete expansion tanks.

- The second and 'conventional' ICCP method for protecting a swimming
  pool uses anodes mounted from the outside (i.e. the dry side of the pool)
  in the corrosion affected area. Anode types used are internal anodes,
  conductive paint, titanium mesh, ribbon or others applied directly to the
  outside of the corrosion affected area. This method can be used instead
  of the water anodes, and is applied from the outside of the pool. In
  Denmark it is normal for internal anodes to be used to protect the parts
  of the reinforced concrete structure that the water anodes cannot reach,
  such as support columns, promenade deck slabs and diving towers.

To control the performance of the cathodic protection system a number of reference electrodes are normally installed. In the water, zinc reference electrodes are
used. A 25 m × 12 m swimming pool would normally require 4–10 zinc reference electrodes. In the concrete, silver/silver chloride reference electrodes are
generally used for absolute potentials and depolarisation and the cheaper mixed
metal oxide reference electrodes are used for depolarisation measurements only.

The use of computer control has been shown to be very advantageous for
cathodic protection systems in swimming pools, especially for water anode
systems. The current requirement can vary significantly in different areas of
the pool, and at different times of the day. A computer control system can take
readings from many reference electrodes into account and adjust the individual
anode outputs to optimise performance on a real time basis.

7.5.3 Description of Rehabilitation Works

Concrete repairs to the pool walls, beams, columns and promenade slab soffit
were carried out using a general purpose, prebagged repair mortar. Pad repairs
were used to increase cover to reinforcement on certain pool wall patches.
Shuttered repairs using flowing prebagged concrete were used on columns that
were very badly damaged. Around the pool the steel stanchions were begin-
ning to deteriorate due to corrosion. These were repaired by gritblasting and
recoating and some were strengthened with the addition of a steel collar.

Once the repairs were complete, the anodes for the cathodic protection
system were installed. The schematic layout for the ICCP system can be seen
in Figure 7.11.

The pool anodes were proprietary modules comprising a monolithic base
and removable lid, both manufactured from XLPE. The twelve anode boxes
were mounted in place, flush with the tiles on the internal face of the pool

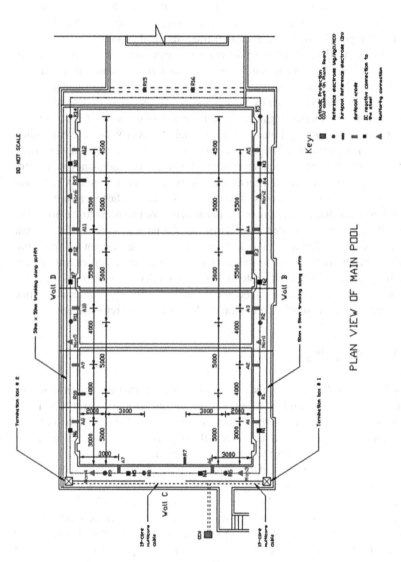

FIGURE 7.11 Anode and reference electrode layout.

around its perimeter. The pool anodes work by using the pool water as the current distributing electrolyte. The current distributor, housed in the XLPE boxes consists of a length of MMO coated expanded titanium mesh. These water anodes were applied to the main pool walls at just below the fill depth. Zinc reference electrodes were fixed in similar, but smaller, XLPE boxes around the pool for dynamically controlling the anode outputs.

Fourteen silver/silver chloride reference electrodes were embedded in holes in the outer walls adjacent to the reinforcement to monitor potentials for controlling the anode outputs and evaluating the system's performance.

Connections were made to the anodes and reference electrodes outside the pool wall in conduit, running around the dry side pool perimeter. Care was taken to ensure that all wiring holes passing through the pool wall were back filled correctly using a water-tight epoxy mortar. DC negative connections were made to the pool wall reinforcement ensuring that a degree of redundancy was maintained. The perimeter conduit carried all DC positive anode wires, DC negative and reference electrode wires back to the transformer rectifier enclosure.

A wall-mounted enclosure in the machinery room adjacent to the pool contains the computer controlled transformer rectifier units that provide DC current to each of the anode zones, and electronics that carry out 'instant off' potential measurements from the reference electrodes. The computer stores transformer rectifier outputs, 'instant off' potentials from the reference electrodes, monthly 24-hour depolarisation data and other data on flash memory. A dedicated phone line and modem allow all information on the system to be retrieved remotely.

7.5.4 Results

The ICCP system was commissioned in stages after final configuration of the software, timer, infrared interruption devices and remote control. The timer is set so in the normal opening hours of the pool the CP is switched off. As insurance that there is no anode current flowing when people are swimming, a movement sensor also cuts the current to the anodes.

The total current demand after operating for a few months was 572 mA at 4.5 V DC. The graphical plots provided in Figure 7.12 shows the steel potentials recorded as 'Instant Off' readings against the Ag/AgCl reference electrodes on the dry side of the pool wall and the zinc reference electrodes on the wet side of the pool wall. Note that the readings for the zinc reference electrodes have been converted to Ag/AgCl readings for clarity. As the wet side of the pool reacts more rapidly to the application of the ICCP, the four zinc reference electrodes were shown on a separate graph.

From the graphs the effect of the approximate five-hour daily application of the ICCP can be seen to polarize the steel. When the pool is in use, the

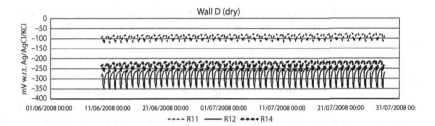

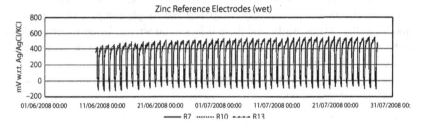

FIGURE 7.12   Showing potentials on the dry and wet sides of the pool respectively.

current is switched off and the potentials decay steadily over the remaining 19-hour period.

The steel in the pool basin is becoming more negative in potential (particularly on reference electrode 16 – the most deeply buried), demonstrating that the steel in this region is becoming, in an electrochemical sense, more reactive or less passivated. The most probable reason for this behaviour is there is insufficient oxygen to reform the oxide layer on the steel. This can be explained by the cathodic reaction using up the oxygen. If sufficient oxygen cannot diffuse through the concrete to re-supply the cathodic reaction then there is an oxygen deficiency, which causes the potential to drop. This does not mean that corrosion is occurring, as oxygen is required in the cathodic reaction, and if no oxygen is present, no corrosion will occur.

Once every month, a 24-hour depolarisation test is also automatically carried out by the computer controlling the cathodic protection system. This test shows more than 100 mV depolarisation on all the steel associated with the reference electrodes.

7.5.5 Conclusions

The swimming pool was a severely corrosion damaged structure with only a limited structural life if temporary repairs had been used, or large financial and operational implications if total replacement had been recommended.

The use of ICCP as a repair technique is now becoming a standard technique in many parts of Europe, but for the UK there is some novelty in protecting the pool basin steel reinforcement using the pool water as electrolyte. The use of water as an electrolyte is common in the conventional cathodic protection industry.

The barriers for using water anodes in swimming pools have previously been the aesthetics and the necessity, for safety and vandalism reasons, to avoid contact between the bathers and the anodes. The installation of the anodes recessed into the walls has overcome these barriers. Additionally, this way of installing the anodes is relatively cheap, fast and simple compared to applying anodes to the outside walls, and there is less to go wrong. Using water anodes reduced both the closure time of the swimming pool and the work in confined spaces giving a considerable cost saving.

The results show that ICCP is being effectively achieved, even after only a short time in operation.

7.6 COAL STORAGE SILOS

7.6.1 Introduction

7.6.1.1 3000T Silos

The two 3000-tonne capacity reinforced concrete coal silos are located in the Hunter Valley, New South Wales, Australia. The silos store coal delivered straight from the Coal Handling Preparation Plant ready for filling rail wagons for transport to the ship loader in the Port of Newcastle, New South Wales. The storage silos are a primary company asset, and a vital infrastructure link to facilitate the efficient delivery of coal.

The silos were originally constructed in the early 1980s using conventional staged construction work practices. Less than 12 months after construction, cracks were noticed on the external faces of the silos and post-tensioned steel cables were externally installed to each silo. The externally applied cables acted to strengthen the walls to overcome the lack of adequate hoop action strength principally due to bulk solids and thermally applied loads.

The strengthening cables had a warranty of ten years with their expected lifespan dependent upon maintaining sheath integrity. During a load-out audit in December 2004 it was noted that cables had failed (see Figure 7.13), thereby posing an immediate safety concern. Silo capacity was reduced to 75% and ongoing capacity reductions presented a serious commercial risk to operations.

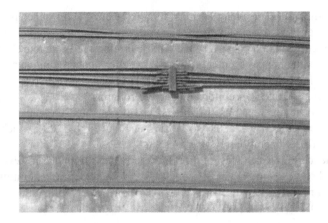

FIGURE 7.13   Failed post-tensioned cables on 3000T silos.

Potential corrosion of reinforcement and concrete deterioration was also evident to both silos.

The original silo design life was 20 years (from 1980), but mine operations now dictated a required life of 50 years to a Life of Mine (LOM) until 2030.

7.6.1.2 5000T Silos

The 5000T silo (see Figure 7.14) was constructed in 1971 and forms part of the coal-handling facilities of another coal mine in the Hunter Valley, New South

FIGURE 7.14   Elevation of 5000T silo.

Wales, Australia. The silo is a critical link in the coal feed from conveyor to a nearby coal-fired power station.

Silo construction consists of a cylindrical reinforced concrete shell. Mild steel forms the hopper section of the silo, which is anchored into a reinforced concrete ring beam above the footing level. The reinforced concrete ring beam is supported by internal reinforced concrete columns.

Previous investigations (by others) had down rated the capacity of the bin to 3700T and it was likely to be further reduced to 3200T which would severely hinder operations.

The original silo design life is not exactly known; however, it is believed to be 25 to 40 years (from 1971). However, mine operations now dictate a required life of 60 years to a Life of Mine (LOM) until approximately 2030.

7.6.2 Condition Investigations

Investigations were carried out from a dedicated man-box suspended from a mobile crane or using elevated work platforms and rope access techniques for each silo, internally and externally. This work included:

- Visual inspection.
- Mapping of cracks.
- Delamination (drummy) survey.
- Electrode potential surveys (at representative locations).
- Concrete sampling for chloride and sulphate profiles.
- Concrete core extraction for compressive strength and cement content determination.
- Concrete breakouts to inspect reinforcement corrosion state.

7.6.3 Investigation Results – 3000T Silos

7.6.3.1 Engineering Review – Structural Analysis/Assessment

The structural analysis/assessment in summary identified that the enhancement of strength offered by the post-tensioned cables was considered paramount and their effect needed to be restored. However, further strengthening was also required due to vertical bending near the silo floor, as well as sway of the silo under temperature and high shear loads causing cracking to the lower dividing shear walls. All were considered to be overcome with the application of high-strength carbon fibre laminates (CFL) bonded to the concrete surface.

7.6.3.2 Field Investigation

Concrete Strength. Compressive strength of test cylinders varied from 34 MPa to 43 MPa. The original design plans nominated a compressive strength of 20 MPa for footings and 25 MPa for walls. Cement content varied from 10% to 12% by mass of concrete (i.e. 240–290 kg/m³ of concrete based on an assumed dry bulk density of concrete of 2,400 kg/m³).

Chloride Analysis. Sampling was conducted on the external faces of the silos and the inside of Silo 2 where substantial internal cracks were evident so as to determine if chlorides were a mechanism of corrosion initiation of wall reinforcement at large internal cracks.

The 3000T silos were very inland (~80–100 km) and remote from sea or saline water masses resulting, as expected, in chloride results being very low (i.e. <0.05% by weight concrete). However, chloride analyses were also carried out to confirm whether any cast-in chloride issues existed.

Carbonation Analysis. Reinforcement corrosion induced damage (delaminations and spalled areas) was evident to the external surfaces of the silos. Concrete cover (by cover meter and tape) and carbonation depth measurements (using phenolphthalein pH indicator solution) were taken at various locations on the silos.

Results indicated that carbonation has occurred at shallow depths (5–20 mm) where significant cover was evident, therefore not posing any widespread corrosion risk to the reinforcement.

Carbonation was found to pose a problem to the silo structures at cracks 0.5 mm wide and greater where carbonation depths had reached the rebar in nearly all cases (Figure 7.15).

The cover to reinforcement was considered high enough not to raise any concern for Life of Mine (LOM) of >20 years (i.e. till Year 2030) except where carbonation-induced corrosion is a mechanism of deterioration at cracks greater than 0.5 mm and at low cover areas.

Electrode Potentials. Electrode potentials were recorded and mapped (mV vs. Ag/AgCl/0.5M KCl) resulting in the most negative readings (i.e. at least −50 to −125 mV) being experienced in isolated locations over cracks; however, most readings became more positive (i.e. at least +50 to +150 mV) within 250 mm of negative readings. This indicated that the corrosion activity was localised at crack locations.

Assessment of Existing Post Tensioned Cables. Deterioration of the external post-tension cables was evaluated by using eddy current techniques and

FIGURE 7.15 Increased carbonation depths at cracked concrete.

visual inspections to determine if any remaining section of the cables presented an immediate safety risk.

Eddy current results were calibrated with visual observations to establish if a cable was either sound or had wires broken. Where more than two wires were broken (see Figure 7.16), the cable was removed prior to the commencement of the works using an elevated work platform (EWP) and clamp plates at the zone of cutting.

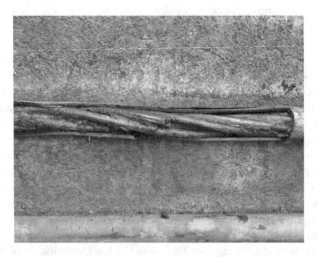

FIGURE 7.16 Corrosion to post-tensioned cables.

7.6.4 Investigation Results – 5000T Silo

7.6.4.1 Engineering Review – Structural Analysis/Assessment

Similar to the 3000T silos, a 3D finite element (FE) model and analysis was undertaken while considering loads imposed from stored coal, earthquakes, wind and thermal effects.
Analysis outcomes included:

- Original design capacity was verified as 5000T.
- Assuming all internal reinforcement is redundant and no remedial works are carried out, 3500T (including hang-up) capacity can be achieved based on assumptions made.
- Assuming all internal reinforcement is redundant and concrete loss is not more than 50 mm, 5000T could be achieved by strengthening with external post-tensioned cables above the ring beam. Cables need to be at least 13 mm diameter (6×25, 1770 MPa) at 650 mm vertical centres. Strengthening is also required for the wall vertical moment capacity, using CFLs, minimum grade of 150/2000 (50/1.4 mm), 8 m long and at 750 mm centres around the silo circumference.

7.6.4.2 Field Investigation

To establish if the bin could be upgraded, a desktop study was undertaken prior to the commencement of a detailed site investigation. Any detailed site investigation would include internal examinations that would be difficult due to an estimated 900T of coal hang-up that had existed for many years. Hang-up had not been previously removed due to the difficulty in gaining access from the top of the silo due to the head house and roof cone arrangement.

Wall Faces. Visual mapping of the internal surface damage revealed that about 50% of the wall had suffered from spalling and exposed reinforcement. Likewise, low cover areas also resulted in spalling externally, although not as severe as the inside wall face.

Site Thickness Surveys. Non-destructive thickness testing equipment was used to estimate the substrate thickness of the silo wall. A calibration of the echo pulse thickness tester was conducted at ground level, i.e. substrate thickness of 200 mm. Once calibrated, a thickness survey was then performed on the silo walls between the heights of 21 m and 27 m; this returned thickness readings between 130 mm and 180 mm. Once these thickness readings were completed, the echo pulse tester was cross-checked at ground level again.

The readings should have returned thicknesses of 200 mm; however, it only returned readings ranging from 160 mm to 180 mm.

The echo pulse thickness tester estimates the concrete thickness by recording the time taken for a sound wave to travel through the concrete medium. Areas of concrete with a less dense medium (i.e. poor compaction, different batches etc.) result in a quicker travel time and hence a smaller thickness recording.

The variation in these readings was such that they could not be used in the design analysis. Assumptions were made (150 mm thick substrate) to allow for the design analysis to be completed.

Chloride Analysis. The level of chlorides both internal and external was less than 0.05% by weight concrete for all depth increments. As for the 3000T silos, the 5000T silo was also very inland (~80–100 km) and very remote from sea or saline water masses resulting, as expected, in chloride results being very low; however, chloride analyses were also undertaken to confirm whether any cast-in chloride issues existed.

Carbonation Analysis. External wall face carbonation depths varied between 20–30 mm. Internal wall face carbonation depths were higher at 30–45 mm. These depths of carbonation corresponded with the depth of low cover horizontal reinforcement. Higher internal carbonation depths have resulted from the concrete not being wetted.

7.6.5 Remediation

7.6.5.1 Options Considered – 3000T Silos

The various remedial options considered for the 3000T silos were as follows:

- Option 1 – Do nothing
  - Failure of existing structure with exposure to potential investigations and delays.
  - Possible damage to other infrastructure due to a major structural failure.
  - Unexpected interruption in operations would result in a delayed delivery of coal to Newcastle Port resulting in unacceptable commercial penalties.
  - Unplanned capital impact on future operational budgets.
- Option 2 – Remediate
  - Failure of existing structure prior to commencement of works.
  - Loss of production while one silo offline during internal repairs.
- Option 3 - New silos at adjacent location

- Failure of existing structure prior to commencement of works.
- Significant additional costs with planning, design, approvals, conveyors etc.
- Ongoing interim silo maintenance until new structures commissioned.
- Option 4 – New silos in-situ
  - Failure of existing structure prior to commencement of works.

Option 2 was the option selected.

7.6.5.2 Options Considered – 5000T Silo

Four remedial options were considered for the 5000T silo and they included:

- Option 1 – Make safe, no remedial works or strengthening, load capacity downgraded (0–10 years' service life).
- Option 2 – Make safe, remedial works and strengthening carried out, load capacity rated to 5000T (0–10 years' service life).
- Option 3 – Make safe, no strengthening, load capacity downgraded (10–20 years' service life).
- Option 4 – Make safe, remedial works and strengthening carried out, load capacity rated to 5000T (10–20 years' service life).

Option 4 was the option selected.

7.6.6 Outline of Adopted Remediation Option – 3000T Silos

The 3000T Silos had undergone remediation strategies for strengthening very soon after original construction. The past strengthening technique (i.e. post-tensioned cables) had reached the end of their serviceable design life. Based on the site investigation and structural assessment, the proposed remedial works for the silos consisted of removal of the existing post-tensioned cables in a predetermined sequence ensuring structural capacity was not compromised, repair of major cracks (0.5 mm and above) or where concrete was deemed to be drummy (or unsound), placement of CFLs (vertical and horizontal) and the application of an elastomeric anti-carbonation coating to the external surface of the silos. Cementitious repair mortar used internally was formulated with a higher than normal tensile strength to minimise the requirement of internal carbon fibre laminates and wraps.

Repairs were also performed on the roof structure, in the form of purlin and sheeting replacement and protective coatings to structural steel.

Cables were removed and replaced with carbon fibre through a managed staging process to minimise energy loss to the silo and to enable works to proceed whilst the silo was used to store a limited volume of coal.

Zones were marked up on the concrete work face to indicate unacceptable crack widths, drummy zones and spalled areas that required repair. Hydro-demolition was used to excavate concrete and prepare exposed reinforcement. Such a technique was not originally accepted by the client; however, after presentations and demonstrations (including safety management) of the process, this view was very quickly altered. After preparation of exposed rebar and application of a rebar coating, proprietary repair mortar (shotcrete) was then installed using a low-pressure spray application. A protective elastomeric acrylic anti-carbonation coating was applied to all exposed silo concrete surfaces.

Carbon Fibre Laminates (CFLs) replaced the removed post-tensioned cables by epoxy-fixing horizontally to the concrete face. CFLs were also used vertically over the horizontal laminates.

7.6.7 Outline of Adopted Remediation Option – 5000T Silo

Investigations all indicated that a lot of reinforcement to both the inside and outside faces of the silo wall had been built with low cover. Due to the difficulty in gaining internal access to the silo it was decided to allow the internal face to deteriorate for a controlled period and carry out repairs and strengthening to the external face. External strengthening would supplement the section loss of up to 50 mm from the internal face as well as all steel reinforcement at the internal face being assumed as structurally redundant. Such a strategy would therefore minimise the extent of internal repairs.

To implement the above strategy, CFLs were required vertically on the outside face in the zone from 2 m below the ring beam for a length of 8 m. Horizontal strength was enhanced by the installation of horizontal post-tensioned cables in the same area.

Concrete repairs entailed breakout by hydro-demolition, rebar preparation, rebar coating application and proprietary repair mortar (shotcrete) installation by low-pressure spray.

An approved anti-carbonation coating was applied to the outside face of the silo.

7.7 MARINE WHARVES

7.7.1 Introduction

Corrosion-induced deterioration of the reinforced concrete wharf assets within the Port of Newcastle (PoN), New South Wales, Australia occurs in the aggressive marine environment, and corrosion management approaches are necessary to sustain the service lives of structural and building assets. For the PoN, some of their wharf reinforced concrete substructure elements have begun to deteriorate within their 50-year design lives and others with ages in excess of 110 years have deteriorated very little.

Various repair and protection technologies and approaches are possible during the corrosion initiation (t_0), metastable pitting/corrosion onset (t_1) and propagation periods (t_2) of chloride affected reinforced concrete structures as indicated at Chapter 6. The PoN have adopted a number of such approaches for select wharf substructure elements to prolong their service life (with consideration of operational requirements and constraints). The approaches adopted have included (but are not limited to) 'do nothing', conventional patch repair, preventative surface treatment (penetrants and coatings) and concrete cathodic protection (impressed current and galvanic).

Impressed current cathodic protection (ICCP) has been the corrosion management approach adopted by the PoN when reinforcement corrosion requires stopping in critical reinforced concrete wharf substructure elements (regardless of the chloride content of the concrete and no matter the rate of corrosion of the steel reinforcement).

7.7.2 Port of Newcastle Wharf and Berth Structures

The wharf and berth structures of the PoN are of reinforced concrete construction (decks, substructure beams and rear walls) supported on reinforced concrete or steel piles. Figure 7.17 shows an aerial view of the PoN.

The PoN West Basin 3 wharf (Figure 7.17) has an ICCP system installed to select substructure reinforced concrete elements. The West Basin 3 CP system to the front beam soffit substructure elements was installed in 1998. Other ICCP systems have been installed to West Basin 4, East Basin 1 and 2 wharves and the Kooragang K2 Wharf between 2002 and 2005 (Figure 7.17).

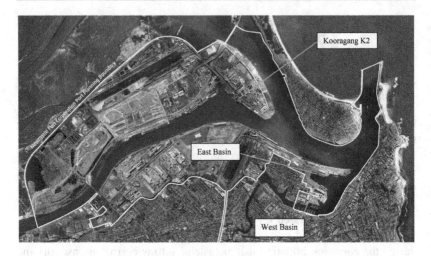

FIGURE 7.17   Aerial view of the Port of Newcastle, New South Wales, Australia.

7.7.3 Concrete CP Systems Overview

Impressed current anode systems utilised by the PoN include catalysed titanium ribbon mesh, mixed metal oxide (MMO) coated ribbon, MMO-coated discrete anodes (tailor-made) and titanium-based ceramic discrete anodes. Transformer rectifier units (TR units) vary in number and type. Remote monitoring and control systems (RMCS) were installed to some TR units. The RMCS units had operational reliability issues from commissioning and no remote monitoring and control of the CP system has been performed (and the RMCS units were discarded long ago). Monitoring has been undertaken by manual means throughout the life of the CP systems. Key aspects of the CP systems and their extent of installation are summarised at Table 7.2.

For substructure beam or deck soffit elements, ribbon mesh and ribbon anodes cementitiously grouted into slots (refer Figure 7.18) or chases cut into the concrete surface have been utilised. At spalled areas the ribbon mesh anodes were suspended from the reinforcement using plastic fixings prior to application of shotcrete (Figures 7.19 and 7.20). The grouts and the shotcretes used were proprietary cementitious and CP compatible with known electrical resistivity characteristics and increased alkalinity (buffering capacity) to resist acidification (since the electrochemical reactions at the anode to grout interface are oxidising, producing acidity).

TABLE 7.2 Port of Newcastle – Concrete CP System Details

STRUCTURE	CHAINAGE (M)	ELEMENTS PROTECTED BY ICCP	ANODE TYPE	TR UNITS		NUMBER OF PERMANENT REFERENCE ELECTRODES	DATE OF SYSTEM COMMISSIONING
				NUMBER	OUTPUTS (CHANNELS) PER TR UNIT		
West Basin No. 3	0–229	Front beam soffit	Ribbon mesh	2	4	20 (TR 1) 18 (TR 2)	November 1998
West Basin No. 3	0–229	Rear deck soffit	Ribbon mesh	2	4	10 (TR 5) 10 (TR 6)	November 2014
West Basin No. 4	229–518	Front beam soffit	Ribbon mesh	2	6 (TR 1) 10 (TR 2)	26 (TR 1) 32 (TR 2)	July 2002
West Basin No. 4	229–518	Rear deck soffit	Ribbon mesh	2	4	10 (TR 7) 10 (TR 8)	November 2014
East Basin No. 1	0–137	Front deck soffit, front beam, rear deck soffit, rear beam, cross-beams	Ribbon	1	10	48	February 2005
		Piles	Ceramic discrete				
East Basin No. 2	137–385	Front deck soffit, front beam, rear deck soffit, rear beam, cross-beams	Ribbon	2	10	48 (TR 1) 48 (TR 2)	October 2004
		Piles	Ceramic discrete				

(Continued)

TABLE 7.2 (Continued) Port of Newcastle – Concrete CP System Details

| STRUCTURE | CHAINAGE (M) | ELEMENTS PROTECTED BY ICCP | ANODE TYPE | TR UNITS | | NUMBER OF PERMANENT REFERENCE ELECTRODES | DATE OF SYSTEM COMMISSIONING |
				NUMBER	OUTPUTS (CHANNELS) PER TR UNIT		
Kooragang No 2	0–230	Front deck soffit, front beam, back beam, rear deck soffit, rear beam, end cross-beams	Ribbon mesh	1	10	50	September 2005
		Piles	Discrete (tailor-made)				

FIGURE 7.18   Ribbon mesh anode installation into slots.

FIGURE 7.19   Ribbon mesh anode suspension from reinforcement at spalled areas.

7.7.4 West Basin Wharf 3 (Front Beam)

7.7.4.1 Background

This concrete ICCP system is the oldest in the Port, commissioning having been completed in November 1998. It protects the front crane beam soffit (width ~2.2 m) for a chainage of 229 m.

FIGURE 7.20   Shotcrete reinstatement over ribbon mesh anode.

The condition assessment/survey works identified:

- Damage is generally confined to the front beam and the rear beam.
- The level of chloride penetration associated with the beams near the front of the wharf (being more exposed to wave action) is considerably higher compared to the rear beams, excepting the very rear beam.
- The soffit and front face of the beams contain higher chloride levels than the more protected rear face. Again, the exception is the rear beam which is exposed to reflected wave action off the sea wall.
- Electrode potential measurements revealed more negative potentials towards the front of the wharf and on the rear beam.
- The more negative readings were indicative of increased corrosion activity as confirmed by visual inspection of reinforcement corrosion state (at concrete breakout locations) and the gradients of potential. The more negative readings were not due to cathodic polarisation effects due to higher concrete moisture content and restricted oxygen access to reinforcement.

Based on a review of the findings from the condition surveys of each facility, PoN determined that the priority at the time was the rehabilitation of West Basin Wharf. Taking into consideration the variability of both condition and service environment, the front (crane) beam of the wharf was considered separate to the rest of the structure (rear beams).

For the purpose of selecting the most cost-effective repair solution it was decided that a lifecycle cost analysis (with sensitivity variations) be carried out for two options: conventional patch repair and impressed current cathodic protection (CP).

Prior to analysing the CP option, further diagnostic testing was undertaken to assess more accurately reinforcement corrosion activity up the sides of the beam. This testing revealed that corrosion of the stirrups (running up the sides of the beam) had initiated up to a height of approximately 150 mm above the soffit. It was therefore considered that CP could be achieved with the placement of anodes along the soffit of the beam only.

The life-cycle cost analysis was based on the discounted cash flow concept. In essence though, it is more truly a discounted cost analysis as all dollar values are cost with no actual revenues. A figure of 7.5% was used as the discount rate. There may be justification for differing values of this rate but as the analysis was only meant as a comparative exercise, it was assumed that 7.5% would suffice. Inflation was not considered in forward estimates of costs. If inflation was used to increase the input costs, the analysis would require double discounting for the resulting discounted costs to be correct. As such, inflation would have no impact on the final comparison. This is peculiar to an analysis where only costs are the inputs.

The outcome of the analysis was that ICCP was clearly the most cost-effective remedial option. Considering both initial and long-term (20-year) costs, ICCP was less expensive.

Restrained by funding, PoN decided to apply CP to a 229 m section of the front beam soffit as a first stage. PoN adopted the Design and Tender approach for the contract.

7.7.4.2 ICCP System Summary

As summarised at Table 7.2, catalysed titanium ribbon mesh anodes have been utilised, two (2) TR units control the system with each TR unit having four (4) channels/outputs meaning there are four (4) anode zones per front beam soffit (8 anode zones in total) and there are 38 reference electrodes (Ag/AgCl/0.5 M KCl) installed to enable monitoring of system performance. Ribbon mesh anode spacing varies between 70 mm at heavily reinforced sections to 300 mm at lightly reinforced sections. The design current density on steel reinforcement and ancillary steelwork embedded in concrete was 20 mA/m^2.

7.7.4.3 ICCP System Monitoring Findings

Potential Decay (i.e. decay/depolarisation over a maximum period of 24 hours) results measured at the various permanent reference electrodes since system commissioning in 1998 are summarised at Figure 7.21. Figure 7.22 presents the Extended Potential Decay (i.e. decay/depolarisation over a maximum

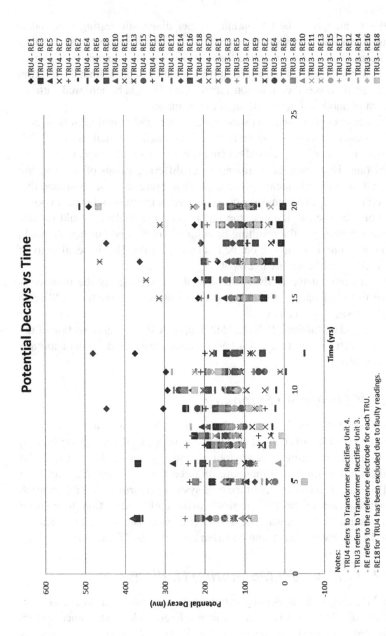

FIGURE 7.21   Potential decay results over time – West Basin Wharf 3 CP system.

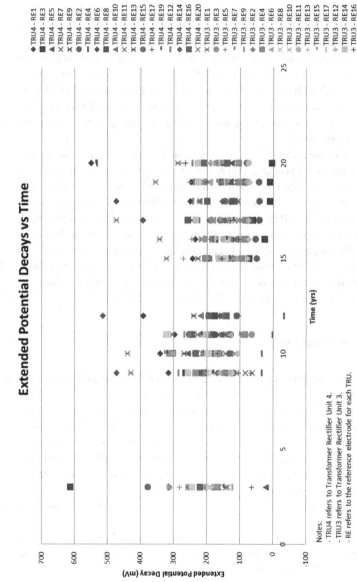

FIGURE 7.22   Extended potential decay results over time – West Basin Wharf 3 CP system.

period of 72 hours) results at the various reference electrodes obtained over the 20+ years of operation of the West Basin Wharf 3 concrete CP system.

The potential decay and extended potential decay results indicate that protection is being provided (>100 mV decay) at the majority of reference electrodes. At those reference electrodes not showing protection, some potential decay is being achieved. Also, with the exception of one (1) reference electrode, at these same reference electrodes either the Absolute Potential criterion or Absolute Passive criterion of AS 2832.5 (2008) is being met. The requirements of AS 2832.5 (2002 and 2008), namely, that compliance with at least one of the protection criteria is necessary and that the compliance shall be maintained on a continuous basis, are therefore not being met at each and every reference electrode.

Where reference electrodes have not shown protection from potential decay or extended potential decay criteria over the last few years, some cathodic current has been received; however, if TR unit output current is increased dramatically for these zones to try to achieve protection, then overprotection occurred in other areas. A practical approach has therefore been adopted whereby the condition of the concrete near these reference electrodes is more closely inspected each year, and conduct of patch repair (local) will be undertaken if and when required in the future. However, at this time (after more than 20 years) no patch repairs have been necessary anywhere within the cathodically protected front beam of West Basin Wharf 3.

7.7.5 Conclusions

The protection criteria of AS 2832.5 have been used to assess performance of the West Basin 3 concrete CP system. After 20 years of operation, CP has been provided to the majority of reinforcement at most (but not all) reference electrode locations. At reference electrodes where protection is not being provided, a practical approach has been adopted whereby the condition of the concrete near these reference electrodes is more closely inspected each year and conducting of localised patch repair will be undertaken if and when required in the future. However, even after 20+ years no patch repairs have been necessary anywhere within the cathodically protected front beam of the West Basin Wharf 3.

Some localised maintenance (e.g. mortar reinstatement, ribbon anode repairs, conduit repair, etc.) has been necessary on some sections of the impressed current CP system during its current 20+ years life.

However, even given the problems with the CP system and even though there will be an ongoing commitment to problem rectification and routine maintenance with the systems in the future, it is felt that if concrete CP had not been applied, then the extent of reinforcement corrosion-induced cracking, delamination and spalling would without question have been much greater and structural and safety concerns would have ensued.

REFERENCES

Chess P, (1993), *Newlands Car Park. –Corrosion Monitoring System, Operations and Maintenance*, February 1993, Wycombe District Council, England.

Dockrill B, Green W and Cooke G (2009), Concrete Coal Silos Assessment, Remediation and Strengthening, Corrosion & Prevention 2009 Conference, Australasian Corrosion Association Inc., Coffs Harbour, NSW, 15–18 November.

Dockrill B J and Green W K (2011), Case Studies of Concrete Repair and Strengthening of Coal Storage Silos in Eastern Australia, Corrosion 2011 Conference, Paper 11015, NACE International, Houston, Texas, USA.

Green W, Linton S, Katen J and Ehsman J (2018), 'Do Nothing' and Patch Repair (Without Anodes) Engineered Maintenance Experiences of Marine Concrete Structures, Corrosion and Prevention 2018 Conference, Australasian Corrosion Association Inc., Adelaide.

Hall C and Hoff W (2012), *Water Transport in Brick Stone and Concrete*, Second Edition, p. 255, Spon press, Abingdon, ISBN 978-0-415-56467-0.

Klinghoffer O and Kofoed B (2018), Application of Impressed Current Cathodic Protection to the Steel End Frames of the Oresund Tunnel, Eurocorr 2018, Krakow, Poland.

Layzell D J, Moore M J, Ali M G and Green W K (1998), Condition Assessment and Maintenance of Wharf Facilities within Newcastle Harbour, 2nd RILEM Conference on Rehabilitation of Structures, Melbourne.

Lunniss R and Baber J (2013), *Immersed Tunnels*, p. 255, Taylor and Francis, Boca Raton, ISBN 978-0-415-45986-0.

Standards Australia (2002) AS 2832.5-2002 Cathodic Protection of Metals - Part 5: Steel in Concrete Structures, Sydney.

Standards Australia (2008) AS 2832.5-2008 Cathodic Protection of Metals - Part 5: Steel in Concrete Structures, Sydney.

Index

Printed in the United States
by Baker & Taylor Publisher Services